Man-made Soils

edited by

W. Groenman-van Waateringe
and M. Robinson

Symposia of the
Association for Environmental Archaeology No. 6

BAR International Series 410
1988

B.A.R.

5, Centremead, Osney Mead, Oxford OX2 0ES, England.

BAR -S410, 1988: 'Man-made Soils'

ISBN 9780860545293 paperback
ISBN 9781407346915 e-book
DOI https://doi.org/10.30861/9780860545293
A catalogue record for this book is available from the British Library
This book is available at www.barpublishing.com

CONTENTS

Page

LIST OF CONTRIBUTORS

Mr. M.J. Allen,
Department of Archaeology, University of Southampton, Highfield, Southampton, SO9 5NH, UK.

Dr. C.C. Bakels,
Instituut vor Prehistorie, Reuvensplaats 4, 2311 BE Leiden, the Netherlands.

Dr. J.A. Bakker,
Albert Egges van Giffen Instituut vor Prae-en Protohistorie (IPP), Singel 453, 1012 WP Amsterdam, the Netherlands.

Drs. R.W. Brandt,
IPP, Singel 453, 1012 WP Amsterdam, the Netherlands.

Dr. C.A.I. French,
3 Fountain Court, Ufford, near Stamford, PE9 3BJ, UK.

Prof. W. Groenman-van Waateringe,
IPP, Singel 453, 1012 WP Amsterdam, the Netherlands.

Dr. H.A. Heidinga,
IPP, Singel 453, 1012 WP Amsterdam, the Netherlands.

Dr. H.C.M. Keeley,
Ancient Monuments Laboratory, Historic Buildings and Monuments Commission, Fortress House, 23 Savile Row, London, W1X 2HE, UK.

Dr. M. Robinson,
University Museum, Parks Road, Oxford, OX1 3PW, UK.

Ir. L.T. Runia,
IPP, Singel 453, 1012 WP Amsterdam, the Netherlands.

Dr. P.C. Vos,
Singel 9, 1012 VC Amsterdam, the Netherlands.

Ir. W. van de Westeringh,
Vakgroep Bodemkunde en Geologie, Postbus 37, 6700 AA Wageningen, the Netherlands.

Mr. T.J. Wilkinson,
26 Alexandra Terrace, Lincoln, LN1 1JE, UK.

1. INTRODUCTION

W. Groenman-van Waateringe and M. Robinson

The existence of raised agricultural soils from the medieval period in the Netherlands was thought to be so intriguing that the AEA was invited to devote the 1985 annual conference to this phenomenon and to man-made soils in general. The meeting place had, of course, to be in the Netherlands. The conference which resulted in this volume took place at Slot Assumburgh, Heemskerk, in the Netherlands, from 20th - 23rd September. The preparation of this volume took more time than at first expected, mainly due to the language problem for Dutch contributors preparing articles for publication in English.

This volume comprises three somewhat separate groups of papers. Firstly, there are papers on plaggen soils. These soils were intentionally raised and have a profile characterised by an unnaturally thick A1 horizon. The paper by van de Westeringh deals with their occurrence in the Netherlands and elsewhere in NW Europe, the kind of material used to raise them and their pedological properties, such as texture, particle size, phosphate content, colour pH and hydrology. Heidinga discusses the conditions which made it necessary to introduce the plaggen system, as illustrated by an example from the central part of the Netherlands. The common view is that plaggen soils came into existence because some means was needed to intensify the production of winter rye. This is challenged by Heidinga, who supposes extreme drought in the 10th century to have been the prime mover for the origin of the plaggen soils, because of their high water retaining capacity. Even after the dry period came to an end, this capacity was of considerable advantage on the dry Pleistocene sandy soils in NW Europe. The articles by Bakels and Groenman-van Waateringe deal with the pollen analysis of the plaggen soils. Plaggen soils are almost unique amongst arable soils because their low pH and high humic content generally result in them being rich in well preserved pollen, including pollen from their crops. The gradual raising of the cultivated surface means that they display a stratified sequence. Bakels discusses the different pollen sources in man-made raised soils and the representation in the pollen record of different crops which could have been grown there. The results are compared with literary sources for crops in the province of North Brabant. The article by Groenman-van Waateringe is a short elaboration of a poster, presented at the conference, and dealing with some arable soils on the Veluwe in the central part of the Netherlands. It discusses the supposed raised soil of 10th century Kootwijk.

The second group brings together articles dealing with various soil changes caused partly by man and partly by natural phenomena. Allen applied magnetic dating, the measurement of magnetic susceptibility and techniques of soil micromorphology as well as more traditional methods to the study of valley colluvium and lynchet deposits in East Sussex. He observes that most late Neolithic/early Bronze Age sites in the valley bottoms of the study area lie obscured beneath plough-induced colluvium. Wilkinson presents a study of manuring practices and their relationship to extensive artefact scatters around some archaeological sites in the Middle East. Surface sampling techniques enabled the definition of lane use zones, while the excavation of soil profiles provided evidence of ploughing, deflation, erosion and sedimentation. A demographic aspect is shown by these artefact scatters which only represent times of peak population. French discusses the nature of the soils in the lower Welland valley, East Anglia, prior to the flooding which gave the Fenlands their present-day appearance. Thin section techniques were applied to palaeosoils underneath Neolithic monuments. They showed argillic brown earth, as develops under stable woodland. Clearance and cultivation from the Neolithic onwards began a process of soil change, including erosion and deposition of wind-blown and colluvial material, only to be halted in low-lying areas by alluviation. Runia describes a study of soils underneath prehistoric burial mounds and questions the phenomenon of so-called secondary podzolization. Micromorphological and chemical analyses do not support the thesis that buried soil profiles act as illuvial horizons for soil formative processes taking place at the present-day soil surface.

The last group of articles is mainly concerned with the effect of soils on human habitation and agricultural practices. The article by Bakker and Groenman-van Waateringe analyses the soils available to the Neolithic Funnel Beaker people and the choices made by them concerning their daily existence (possibilities for agriculture) and their ideology (where to build the funerary monuments). Megalithic tomb (hunebed) construction took place on dry and loamless coversands, mostly in the vicinity of stonefields, where large boulders were easily available. The arable fields were situated on loamy fine coversands. Keely's article deals with her research on abandoned agricultural terraces related to pre-Hispanic settlements in a marginal high desert area of Chile, where, without an irrigation system, agriculture would not have been possible. Evidence for erosion of the terraces, both by water and wind, showed the need for the conservation of soils. Soil properties, such as particle size, organic matter content, pH and nutrient levels are described. Brandt and Vos describe the close relationship between geological development and habitation in the coastal areas of the Netherlands, based on a geological and archaeological survey of three areas in the former IJ-estuary. The result is a detailed landscape reconstruction, combined with settlement history. Various surveying methods were developed for grassland on different soil types, each area requiring its own strategy.

The general conclusion of the conference and the papers presented here is to emphasise the close link between man and soil development. Although soil formation s.s. can be described as an entirely natural phenomenon in terms of chemical and physical processes, the human aspect, especially since the beginnings of agriculture, is ubiquitous on a macro- and micromorphological scale. The awareness of these human-induced processes has wide implications for a better understanding of settlement archaeology. In this respect, however, it is not necessary to think in terms of man being entirely dependent on the environment. The conference gave numerous examples of man not submitting to the restrictions imposed by soil types and climate, but instead actively creating entirely new types of soil. Man is also highly capable of responding adequately to landscape changes induced by his own actions, such as erosion, colluviation and alluvial deposition. The pedologist and archaeologist alike should be grateful to his predecessors for the heritage left to him in the form of buried soils, a palimpsest of evidence for pre-settlement climatic conditions and of subsequent human behaviour.

2. MAN-MADE SOILS IN THE NETHERLANDS, ESPECIALLY IN SANDY AREAS ("PLAGGEN SOILS")

W. van de Westeringh

Abstract

The plaggen soils in the sandy areas of the Netherlands are anthropogenic, characterized by a humic dark topsoil more than 50cm thick (A1 horizon). Cereal cultivation was only possible in a well-balanced agricultural system which included, besides arable land, pastures for the dung producing animals, plus a large area where litter could be obtained. This sytem, existing for centuries, was entirely dependent on natural conditions, i.e. character of the soil and hydrology. To keep the fertility of the poor Pleistocene sandy soils, in use as arable, at the required intensity, it was necessary to add nutrients, i.e. stable muck or dung-laden earth. Eventually, new soils formed on top of the old ones as a result of the inorganic particles in the dung. Plaggen soils show characteristic features. They occur also in other NW European countries, sometimes quite similar to, sometimes differing from the Dutch plaggen soils.

2.1 Introduction

Plaggen soils[1] are wide spread in the Pleistocene sandy areas of the Netherlands (fig. 2.1), although occurring only under specific conditions. Whereas in the Netherlands, under natural soil conditions the A1 horizon and the plough layer (Ap horizon) practically never exceed 30cm, the plaggen soils are characterized by dark, humic top layers (A1 or Ah horizons) of more than 50cm thickness, indicated as Aan horizons ('an' of anthropogenic). Together with plaggen-like soils with A1 (Aan) horizons of 30-50cm thickness they cover about 30% of the Dutch Pleistocene sandy soils (Pape 1970). Plaggen soils are rarely found on loam or clay. If man-made soils (old arable or settlement soils) occur there, they are usually different from those on sand. The genesis of sandy plaggen soils is determined by the natural soil conditions, the hydrology and an agricultural system practised for centuries.

The geology, pedological and hydrological conditions, past farming practices, manuring and soil fertility, the genesis of the plaggen soils and their characteristics, and their occurrence outside the Netherlands will be discussed.

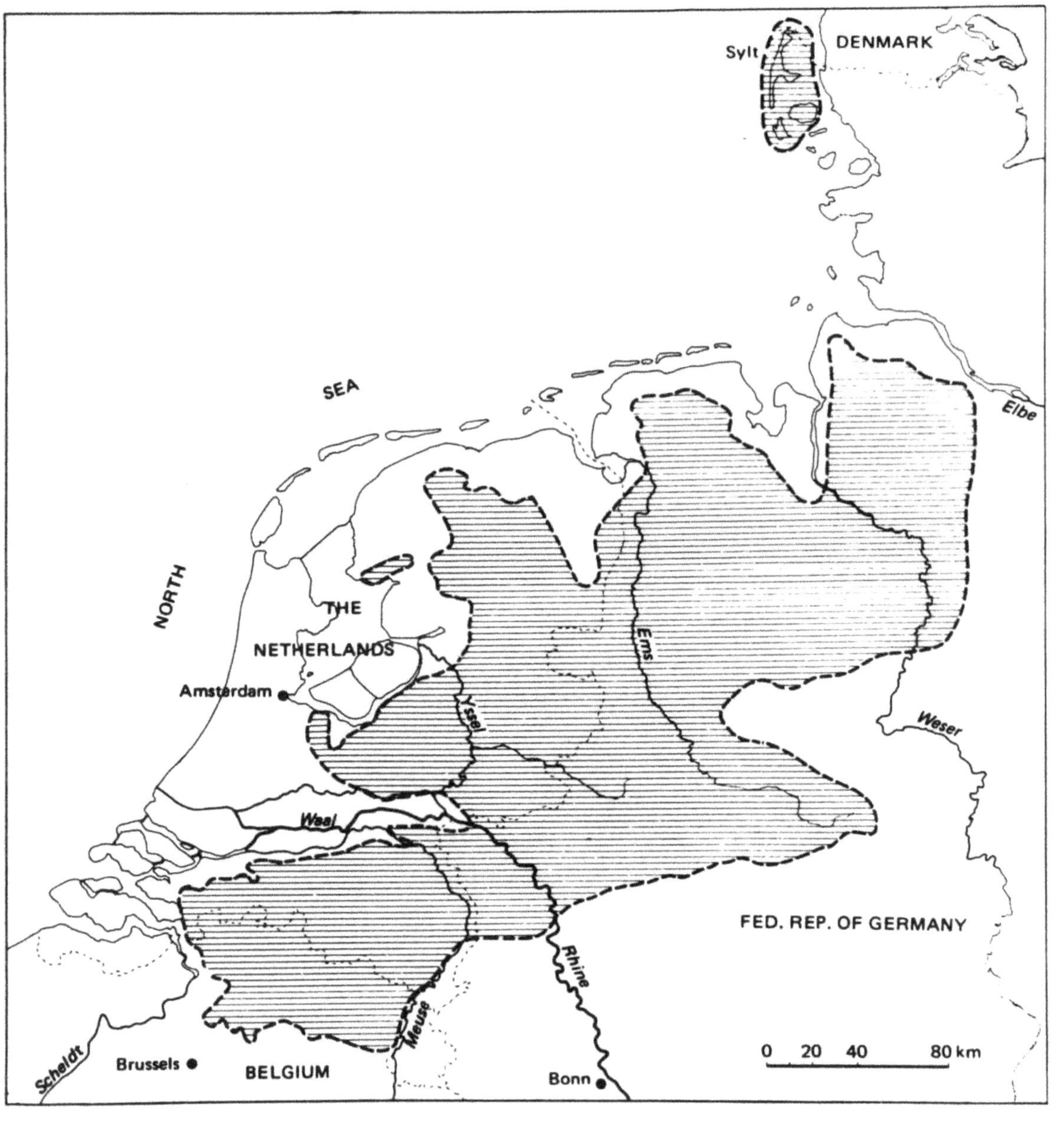

Fig. 2.1 Distribution of plaggen soils in northwestern Europe. (From Pape 1970)

2.2 Geology

The structure of the Dutch landscape has mainly been shaped by geological processes in the last two Glacial periods (van Staalduinen et al. 1979) (fig. 2.2). In the Saale Glaciation, deep valleys were carved out by glaciers, while weight and pressure of the land-ice pushed underlying deposits upwards (pushed ridges). After the retreat of the ice, moraine material (till, boulder clay) of north European origin remained behind in the northern and eastern parts of the Netherlands. In the following Interglacial and the Weichsel Glaciation severe erosion took place of the pushed ridges and deep erosion valleys because of their easily erodable loose material. The most important process of the Weichsel Glaciation in the Netherlands was the deposition of coversand, and, especially in the southern part of the country, of loess. This resulted over large areas in pedogenesis involving parent material with a high quartz content (sometimes more than 90%) and thus a very low natural soil fertility. Only in those places where sand from river valleys was blown in, did soil with a higher fertility form.

Thus we can say that the Pleistocene part of the Netherlands is characterized by low hills and shallow valleys covered for a large part with poor quartz-rich sand, while the rest shows mainly a flat landscape.

2.3 Soils

Dry and poor coversands are characterized by humus podzols (Haplohumods, Soil Survey Staff 1975), wet humus podzols in coversand are classified as Haplaquods. Soils formed in sands blown from river valleys are often classified as acid brown forest soils and brown podzolic soils.[2] The development of the landscape pattern, however, depended on the geogenesis and the extent of each soil type. Moreover, relief and the hydrological situation determined which soils were used for agriculture. Thus in the Pleistocene area because of differences in the nature of the soils, their extent and the number and size of the valleys, prospects for agriculture differ between the NE part of the Netherlands, (the province of Drenthe), those in the central and eastern parts (the province of Gelderland), and those in the southern part (the province of Brabant). Acid brown forest soils are better suited for permanent use as arable than humus podzols. Low-lying humic gley soils make better grasslands than groundwater podzols. The arable, however, could only occupy a very limited part of the whole territory, because the largest part was needed for other purposes (Edelman 1950; van de Westeringh 1979). This will be discussed in the next sections.

2.4 Farming practices in the past

Permanent farming in the Pleistocene sandy areas of the Netherlands was not possible without regular manuring, because of the rather low natural soil fertility of the coversands. Not only the arable, but also the grassland needed manuring under these poor soil

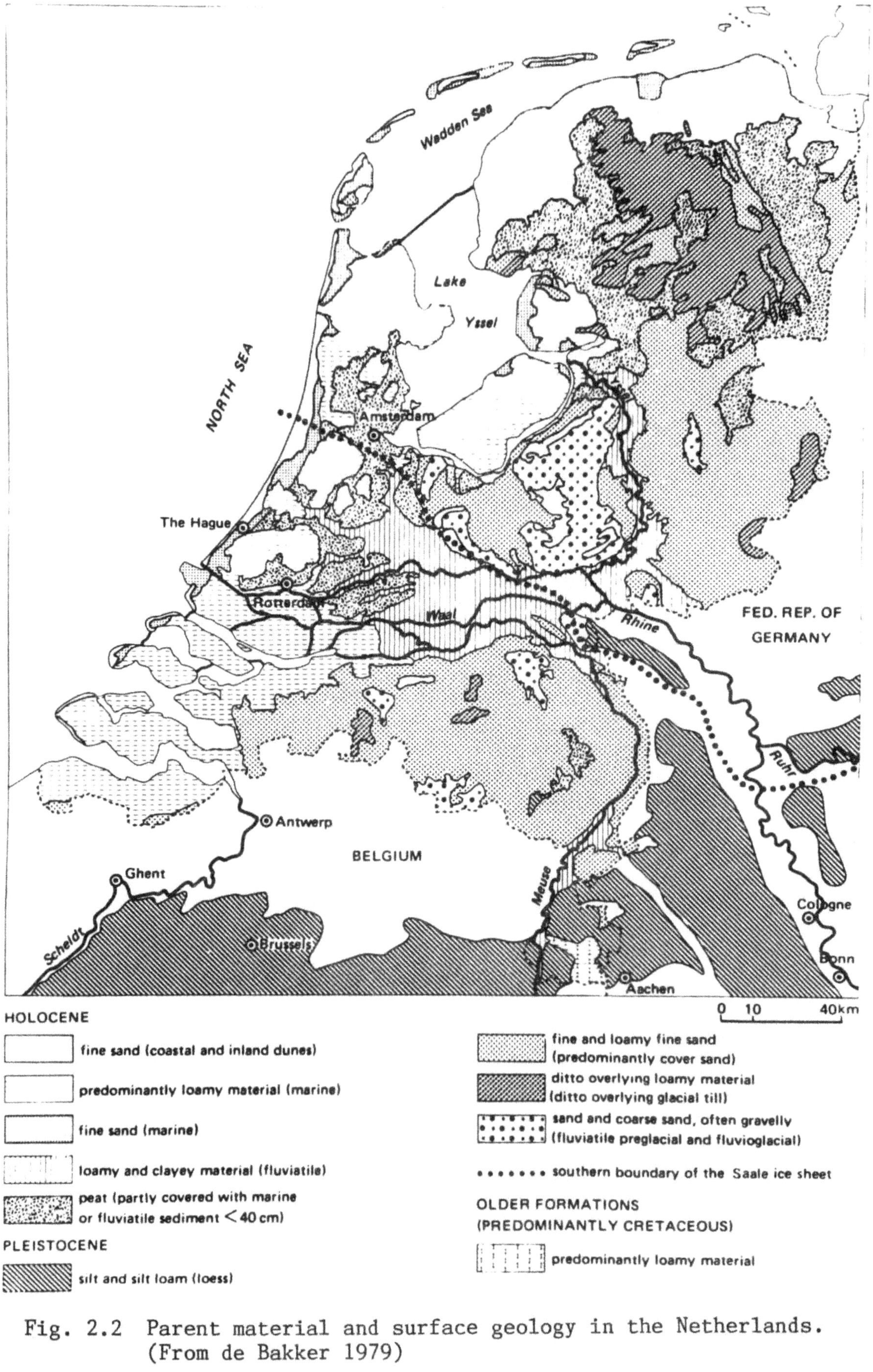

Fig. 2.2 Parent material and surface geology in the Netherlands. (From de Bakker 1979)

conditions. Only in those parts of the Netherlands where fertile soils, such as river clays and marine deposits are present, did the situation differ. An ever increasing population, albeit with fluctuations, since the Neolithic, required more agricultural produce and thus more intensive use of the available arable. On the one hand nutrients are carried away by harvesting the crops, on the other hand they are washed down, out of the root zone, in the wet Dutch winter when rainfall is higher than evaporation. The most economic way of preparing the manure was to gather cow and sheep dung during the winter stabling and/or in the summer during the night. The floors of the stables were strewn with for example forest litter, heather turves and grass sods to absorb the droppings of the animals (Pape 1970) (see fig. 2.3). With the cutting of turves and sods it was inevitable that inorganic material was brought into the stable too. The mineral fraction of the bedding material, especially if it consisted of loam or clay from the fertile grasslands in the brook valleys, provided an extra source of nutrients.

In such a farming system animal husbandry is indispensable and this means that there must be enough pasture land for the livestock and enough waste land for the collection of litter and turves. The ratios between waste land, pasture and arable, dependent as they are, both qualitatively and quantitatively, on the local soil conditions, may differ from one hamlet to another (Table 1, van de Westeringh 1979; van Dis and Rogaar 1986).

Table 2.1 Plaggen soils (ha) and farms (1859) in four hamlets in the municipality of Winterswijk, the Netherlands

Hamlet	Total area (ha)	Area of plaggen soils (ha)	Ratio	No. of farms 1859	Mean area of plaggen soils (ha) per farm
Meddo	2825	378	1:7.47	159	2.40
Ratum	1340	210	1:6.38	73	2.88
Brinkheurne	424	106	1:3.91	42	2.52
Woold	2404	383	1:6.28	149	2.57

The introduction of chemical fertilizers at the end of the last century put a stop to an agricultural practice that went on for centuries.

2.5 Manuring and soil fertility

This age long practise of manuring with litter, turves and sods soaked with dung resulted in the genesis of entirely new, man-made soils with a higher fertility than the original ones. However, the cutting of heather and grass sods diminished the soil fertility of these waste lands. This was especially the case for the humus podzols under the heath, though much less so for the brook valleys where new

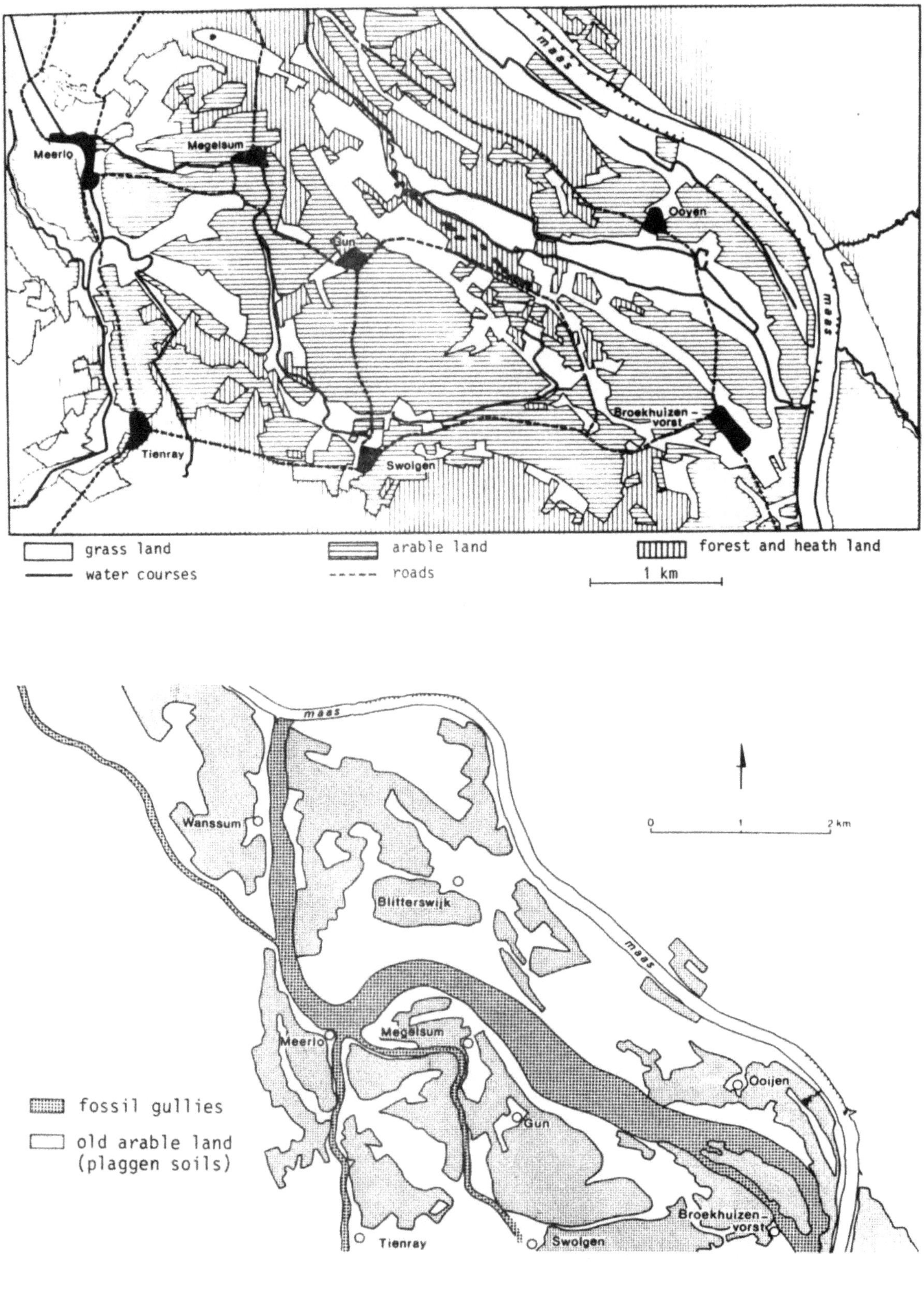

Fig. 2.3 Land use at the end of the 19th century (a) and simplified soil map (b) of an area along the river Meuse, the Netherlands

alluvial deposition compensated for the loss of minerals. The continuing displacement of material from waste land to arable, from poor soils to relatively rich ones, was a heavy burden on the environment and allowed only very restricted arable.

The quality of the manure depended on the origin of the used bedding material and on the kind of livestock. Heather turves, with a high C/N quotient, added only a few elements and thus this kind of manure contained mainly nutrients released through dung decomposition. The colour of the arable soil was influenced by the material added, thus turves with dark-grey to black humus resulted in soils of a similar colour.

Bedding material consisting of grass sods out of brook valleys had a completely different composition. The quality of the humus after decomposition was much better because the C/N ratio was low and more plant nutrients were released. Moreover, the mineral parts (loam or clay) of the grass sods provided the manure with an extra quality. The colour of this kind of man-made soil is brownish, because of a rather high iron content (Knibbe 1969).

On the clayey Holocene sediments old arable soils have deepened A1 horizons which are clearly distinct from the typical plaggen soils on the sandy Pleistocene formations, with their raised A1 horizons. This difference resulted from the higher natural fertility of the clay soils whose higher pH favoured faunal homogenisation.

2.6 Genesis of plaggen soils

The most important property of plaggen soils is their earth content, i.e. mineral particles of sand, clay or loam. After decomposition of the organic matter the mineral components remained on the surface of the fields. Heather turves added sand, whereas grass sods from brook valleys brought loamy or clayey material.

This created raised fields, often with a considerable depth of soil, sometimes exceeding 1m. The original soil profile was buried and a new soil, entirely homogenized by ploughing, was formed. However, the depth of the plaggen soils differs from one area to another. In the northern part of the country, the province of Drenthe, they are scarcely 0.50m deep, which is the minimum thickness for a real plaggen soil in the Dutch soil survey system; in the centre of the Netherlands the depth can be more than 1.20m.

The growth in depth, however, was not uniform over the years (fig. 2.4). In some periods manuring of the arable was much heavier and more material was brought up, in others, fields could lie fallow for some time or were used as grassland. Liter from forests added less mineral material than did turves, while peat or straw added nothing. Nevertheless it has been found that in many places, deeper plaggen soils are older than shallower ones. It seems that in the last phase of the plaggen farming system, the accumulation took place much more rapidly because there was an unsatisfied demand for arable land which was therefore used more intensively, while the amount of waste land available decreased. This resulted in the cutting of thicker turves

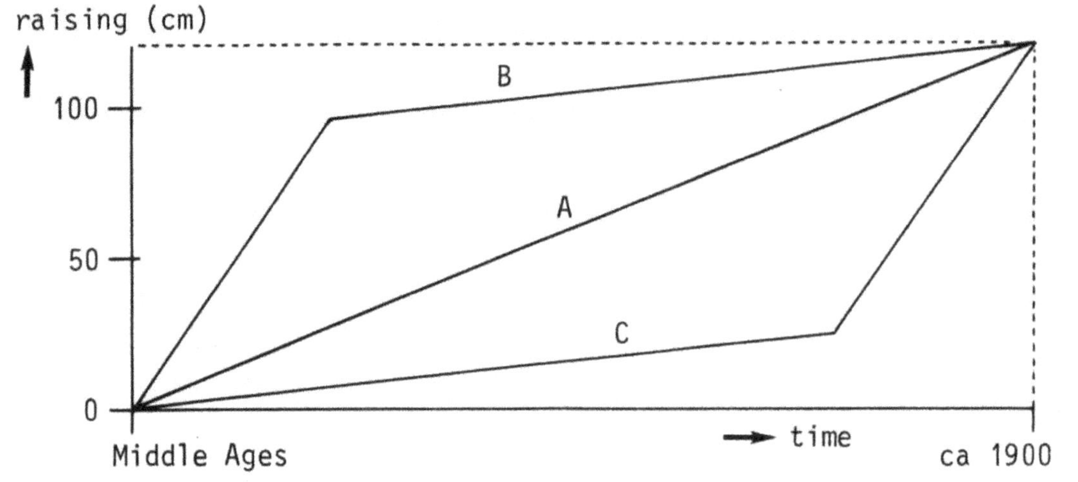

Fig. 2.4 Different possibilities for the accumulation rate of the humic layers of plaggen soils

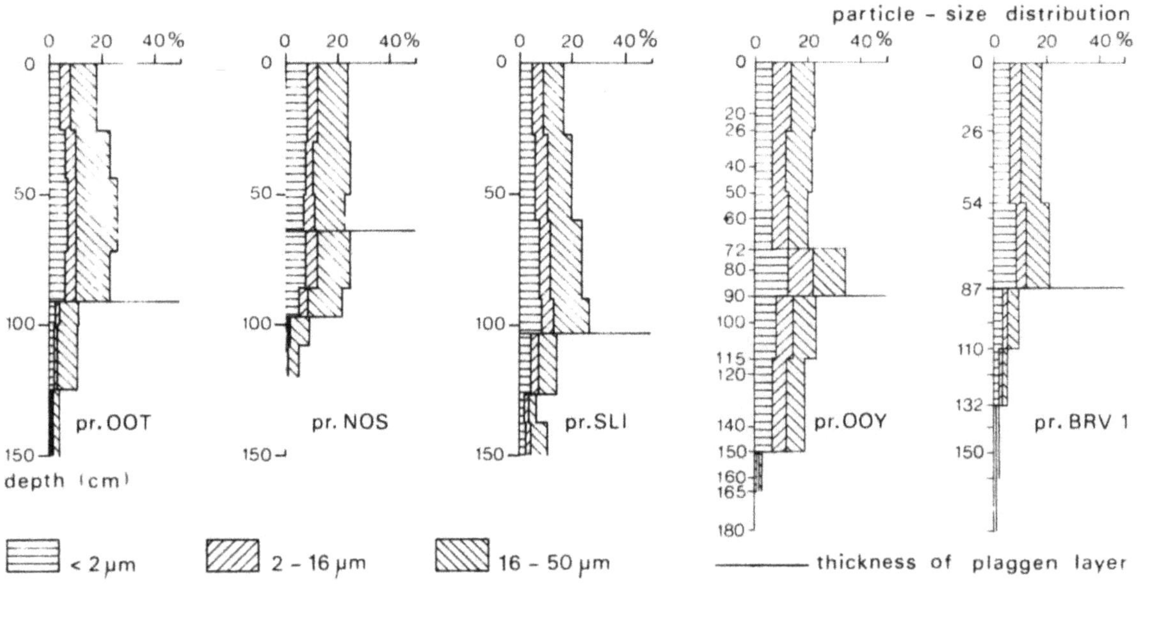

Fig. 2.5 Particle size distribution of five plaggen soils

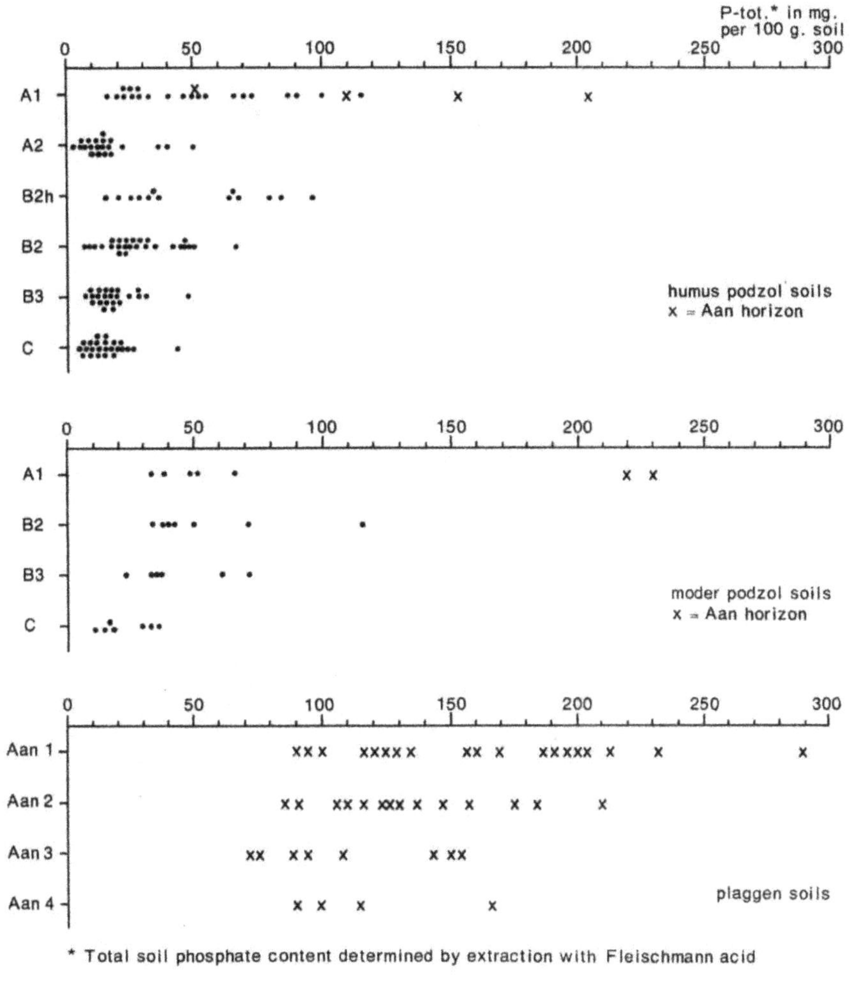

Fig. 2.6 P-total in various horizons of podzol soils and plaggen soils. (From Pape 1970)

with less humus and more mineral material.

2.7 Characteristics of plaggen soils

The particle size distribution of plaggen soils still more or less reflects the origin of the bedding material. In the same area one can find plaggen soils of entirely different textures, ranging from coarse sand to loamy sand or sandy loam with a clay content of 8-10% (fig. 2.5) (van de Westeringh 1970, 1973).

The colour of the plaggen soils varies from dark grey or black to brown (see section 5). In the Dutch soil classification (de Bakker and Schelling 1966), a brown plaggen soil is characterized by colours with a sum of chroma and value more than 4.5 with hues 10 YR and 7.5 YR, or 4 with a hue of 5 YR or redder, together with chromas more than 1. Darker plaggen soils are classified as black.

Compared with other sandy soils in the Pleistocene area, plaggen soils have a high level of water available for the plant, although they lack, as do the other Pleistocene sandy soils, capillary water and groundwater in the root zone (Pape 1970).

Plaggen soils have a low pH and pollen is well preserved in them. Palynological analysis can sometimes reveal the origin of the turves, sods or litter (van der Hammen and Bakker 1971; Behre 1980; Bakels this volume; Groenman-van Waateringe this volume). In black plaggen soils the origin of humus and inorganic material can be traced back to heather, while pollen plants which grew in a moist and fertile milieu can be found in brown plaggen soils.

Climatic conditions in the Netherlands do not permit the identification in the soils of the original chemical composition of the manure, because potassium and nitrates are washed out. Only phosphate fixated as practically insoluble iron phosphate can be found in analyses. Plaggen soils show more total phosphate than other sandy soils in the same area (fig. 2.6).

The homogeneity of plaggen soils, both vertically and horizontally is very important. This supports a crop of uniform quality.

Another important quality is the depth of rooting. Plaggen soils are loose soils with a low resistance and plant roots can penetrate deeply. Moreover, biopores are distributed throughout the entire plaggen soil (fig. 2.7) (van de Westeringh 1970, 1973).

The anthropogenic origin of the plaggen soils usually shows in the presence of artefacts such as charcoal, coal particles, brick and pottery fragments, burned soil particles.

2.8 Plaggen soils in other European countries

The plaggen soils in Belgium and West-Germany with a similar genesis as the Dutch ones will not be discussed here (see Pape 1970; de Bakker 1979). Plaggen soils are not found in the central and northern

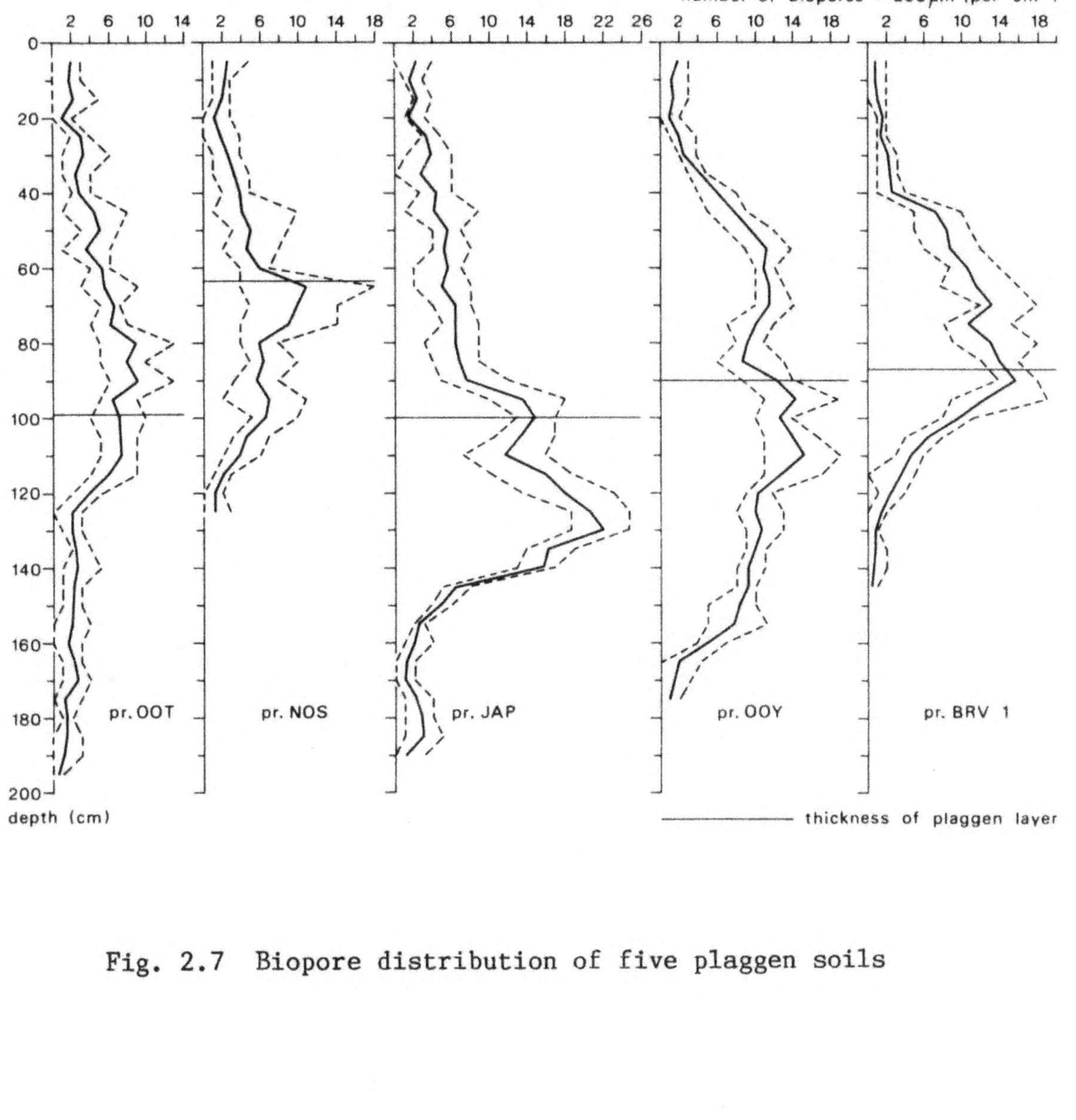

Fig. 2.7 Biopore distribution of five plaggen soils

parts of East- nor probably West-Germany, where more loamy soils, developed from moraine material, are present. It is known, however, from written sources that heather turves were cut and used for the preparation of earthy manure (pers. comm. Dr. Max Linke), although plaggen soils have not been found by the surveys. The reason for this might be, that, because of the higher soil fertility of these loamy soils and as a consequence the larger area that could be used as arable there was not enough stable manure for all the fields, or that the mineral content of the manure was much less. Consequently it is not found in the soil profiles and did not lead to raised man-made soils.

In the Pleistocene landscape of Jutland, Denmark, plaggen-like soils can be found in small fields around villages, more or less comparable to the Dutch plaggen soils, but not so prominent. On Seeland, Denmark, old arable soils on loam are found, not, however, with a raised upper layer (Aan horizon), but with a deepened A1 horizon, caused by biological homogenisation, as was the case in the Dutch Holocene area on the clay soils (see section 2.5).

In Norway one can sometimes find near old settlements alongside fiords small fields with a humus podzol formed on granite but with an A1 horizon that is abnormally thick. This thickness is the result of manuring the fields or garden plots with composted domestic refuse, fish remains, seaweed etc. (Kaland 1986). In some old Dutch fishing-villages, for example on the former island of Wieringen or the Frisian islands this was practically the only available bedding material for absorbing the dung.

Conry (1969, 1970) describes for the Irish Republic so-called sanded soils. In some coastal areas calcareous sea-sand was brought onto the arable in manure, or it was used for sanding the heavy clay soils. Particle sizes characteristic for sea-sand can be found besides the quite different texture of the original soil profile. The Irish raised soils have, in contrast to the Dutch ones, colours matching those of the buried soils and not of the material brought up from elsewhere.

In the Insch valley, Scotland, man-made soils in infields have been described by Glentworth (1954). The depth of the fragipan in the subsoil of the original soil profile compared to its depth in the subsoil of the infields is an indication that the A1 horizon, 80cm thick, is not a completely new, raised layer, but part of the natural soil profile, generated partly by turning over the original soil profile and partly by the mineral particles in the earth-containing manure. These soils are well comparable to the Dutch man-made soils with deepened A1 horizon in clay and loamy soils.

2.9 Conclusions

The genesis of plaggen soils (raised man-made soils used for arable) is closely related to an extinct, historical, agricultural system that for centuries permitted cultivation on the poor sandy soils of the Netherlands. Agriculture at that time was highly dependent on the natural soil conditions. This makes integrated research of landscape and soil, agrarian history, historical geography and archaeology necessary. A thorough knowledge of agriculture in general and the requirements of the different crop plants in respect of the soil and their nutrients, together with soil science (soil suitability, land evaluation) is also needed. Only then will it be possible successfully to study landscape genesis and regional history. Location, genesis and characteristics of plaggen soils can be identified and understood with the help of pedology and agricultural research.

Notes

1. USA Plaggepts, Germany Plaggenböden or Plaggenesche, Belgium plaggengronden, the Netherlands enkeerdgronden.

2. Dystrochrepts or Haplumbrepts when they have a texture finer than very fine sand, loamy very fine sand or finer; Orthents or Psamments if they have a coarser structure.

References

Bakels, C.C., 1988. Pollen from plaggen soils in the province of North Brabant, the Netherlands. In: W. Groenman-van Waateringe and M. Robinson (eds.), Man-made soils. (This volume), 35-54.

Bakker, H. de, 1979. Major soils and soil regions in the Netherlands. The Hague: Junk Publishers, Wageningen: Pudoc. Bakker, H. de and Schelling, J., 1966. Systeem van bodemclassificatie voor Nederland. Wageningen: Pudoc.

Behre, K.-E., 1980. Zur mittelalterlichen Plaggenwirtschaft in Nordwestdeutschland und angrenzenden Gebieten nach botanischen Untersuchungen. In: H. Beck, B. Denecke, H. Jankuhn (eds.), Untersuchungen zur eisenzeitlichen und frühmittelalterlichen Flur in Mitteleuropa und ihrer Nutzung, Tl. II. Abhandlungen der Akademie der Wissenschaften in Göttingen, Phil.-Hist. Klasse, 3. Folge, 116, 30-44.

Conry, M.J., 1969. Man's role in soil profile modification and formation in Ireland. Dublin: thesis.

Conry, M.J., 1970. Plaggen soils in Ireland. Pédologie 19, 321-329.

Dis, A. van and Rogaar, H., 1986. Man-made soils in the Pleistocene sandy landscape. In: Guidebook tour B: soils and landscapes around the Rhine valley, Germany and the Netherlands; XIII Congress of the ISSS. Mitteilungen der Deutschen Bodenkundlichen Gessellschaft 47, 211-223.

Edelman, C.H., 1950. Soils of the Netherlands. Amsterdam: North Holland Publ. Comp.

Glentworth, R., 1954. The soils of the country round Banff, Huntley and Turriff. Edinburgh: Her Majesty's Stationery Office.

Groenman-van Waateringe, W., 1988. Palynology of plaggen soils on the Veluwe, Central Netherlands. In: W. Groenman van Waateringe and M. Robinson (eds.), Man-made soils. (This volume), 55-65.

Hammen, T. van der and Bakker, J.A., 1971. Former vegetation, landscape and man in the Dinkel valley. In: T. van der Hammen and T.A. Wijmstra (eds.), The Upper Quaternary of the Dinkel valley (Twente, Eastern Overijssel, the Netherlands). Mededelingen Rijks Geologische Dienst, N.S., 22, 147-158.

Kaland, P.E., 1986. The origin and management of Norwegian coastal heaths as reflected by pollen analysis. In: K.E. Behre (ed.), Anthropogenic indicators in pollen diagrams, 19-36. Rotterdam: Balkema.

Knibbe, M., 1969. Gleygronden in het dekzandgebied van Salland (with summary: Coversand gley soils in Salland, the Netherlands). Wageningen: thesis.

Pape, J.C., 1970. Plaggen soils in the Netherlands. Geoderma 4, 229-255.

Soil Survey Staff, 1975. Soil taxonomy, a basic system of soil classification for making and interpreting soil surveys. Washington: U.S. Govt. Printing Office.

Staalduinen, C.J. et. al., 1979. The geology of the Netherlands. Mededelingen Rijks Geologische Dienst 31, 9-49.

Westeringh, W. van de, 1970. De opbouw van enige essen bij Denekamp. Landbouwkundig tijdschrift 82, 283-288.

Westeringh, W. van de, 1973. Bodemkundig onderzoek aan oudbouwlandgronden in Noord-Limburg. Landbouwkundig tijdschrift 85, 195-200.

Westeringh, W. van de, 1979. De ontwikkeling van het agrarisch cultuurlandschap in een zandgebied (Winterswijk). Cultuurtechnisch tijdschrift 18, 199-210.

3. CLIMATE AND PLAGGEN SOILS

H.A. Heidinga

Abstract

On the sandy soils of Continental NW Europe a method of soil enrichment has been in operation since the Middle Ages which has resulted in the formation of a characteristic type of man-made soil: the plaggen soil. Here we attempt to assess whether economic and demographic or physical and climatological factors can be isolated which may have caused or stimulated the introduction of plaggen manuring since its adoption represents a major change in current farming practice. Analysis of the available evidence suggested the cause to lie in the extreme climatological conditions which must have affected a great area of Europe during part of the 10th century, i.e. the very period in which plaggen manuring was adopted on a large scale.

3.1 Introduction

Until the extensive introduction of artificial fertilizers towards the end of the last century, the traditional method of restoring fertility to the poor sandy soils of NW Germany, the Netherlands and N Belgium was to spread a mixture of animal manure and cut heather sods or other such absorbent, usually humic, material over the fields. The result of this regular, sometimes even annual, addition of mineral-rich bulk material was the creation of an anthropogenic soil, sometimes reaching a depth of 1.5m: the so-called plaggen soil. Elsewhere in this volume, van de Westeringh describes the characteristics of this particular type of man-made soil, while Groenman-van Waateringe examines the palynological content of some plaggen soils from the Veluwe, a region in the central Netherlands. Here, however, the problem to be tackled is why this labour intensive system was introduced over so wide a region at a particular period in the Middle Ages and why the geographical distribution of the plaggen soils is, in the main, restricted to the above-mentioned area of Europe. It should, however, be born in mind that this system is not exclusive to the Middle Ages, nor to this geographical area: sod fertilization is also attested incidentally in earlier times and in other regions (cf. van de Westeringh, this volume; Kossack et al., 1974).

Fertilization by means of sods is an intensive method of soil enrichment. This is, for instance, brought out clearly in the reminiscences of Harm Tiesing (born 1853), a farmer from Drenthe, who had seen sod manuring still in operation in his youth (Tiesing 1974). The sods were stripped from the heath, far from home, and then had to be transported back to the farm on a cart. Here the sods could either be used as stable litter to retain the animal droppings and urine, or else they might be mixed in with the dung later. When the mixture was well rotted it was brought out to the fields for spreading. Although it is possible that medieval sod manuring differed in some respects, whichever system was used, it was certainly just as labour intensive and was also hampered by high transport costs. According to Slicher van Bath (1963), additional horses were required to maintain the system, which, indeed, would have been inconceivable without the availability of horse traction. Furthermore, extensive tracts of waste land from which to strip the turf were also required; land which was in fact simply denuded. The inescapable consequences were soon manifest, but the degeneration of unclaimed land is unlikely to have prevented farmers from adopting this system. Even today, long term planning is rare enough in the agricultural sector. It is rather the high labour and transport costs which would form the major threshold. In other words, there must have been extremely pressing reasons why the traditional agricultural regime was abandoned in favour of a costly innovation. Although the step to a more productive but also a more labour intensive form of farming is not an invariable concomitant of dire crisis (cf. Boserup 1965), it is likely that sod manuring was adopted as a method of rectifying an increasing shortfall in yields. What needs to be investigated, therefore, is whether the introduction of this system coincided with an agricultural crisis in the respective regions. In order to make the causual association between the two credible, both the nature and the causes of this crisis must be identified and the peculiar advantages of this particular system of fertilization need to be defined.

3.2 Push and pull factors in the introduction of sod manuring

Sub-recent sod manuring was especially prevalent in combination with intensive rye cultivation, whereby a part of the farmland would be fertilized to permit an annual crop (Slicher van Bath 1963). That the association with rye cultivation existed right from the outset is assumed by Behre (1976) on the basis of palynological analysis of a plaggen soil at Dunum (Lower Saxony). Here, the inception of plaggen fertilization on an already well established field system was coupled with an explosive increase in rye pollen. Why, however, this form of manuring should be particularly suited to rye cultivation is not explained. In Dunum there are indications that in addition to the shift from barley and oats to rye, cultivation as a whole was intensified (though this is not to say that the entire cereal crop was increased). The question remains as to what part the material which was added to animal manure played in this process. It is generally assumed that heather sods etc. served both as a medium to retain as much of the stall droppings as possible (van de Westeringh 1978-9) and as an independent measure to improve the structure and fertility of the soil itself. However, research carried out by Pape (1970) indicates that the turves themselves - which were usually stripped from relatively impoverished soils - contribute little or nothing to the chemical fertility of the fields. What does distinguish plaggen soils from non-raised soils in a positive sense is the much improved regime of moisture retention and the more favourable conditions for root development. The moisture available for growth is apparently increased, which is of considerable importance on the high sandy soils, where drought forms a recurrent agricultural problem. The Netherlands Soil Survey Institute classifies the pervious sandy soils with deep natural water table as low grade. Due to moisture shortage in the growing season they are only moderately suited to rye as a winter cereal and are, indeed, too dry for oats or barley as summer cereals. Those areas covered by plaggen soils, which are much less susceptible to drought, are however, more favourably assessed. In this context it should be stressed that only soils at least 50cm deep are classified as plaggen soils (enkeerdgronden).

Raising the level of the fields with soil brought from elsewhere might in fact confer still other advantages. Even waterlogged soils and soils with poor chemical or physical properties may be prepared for cultivation in this manner. Although large areas of poor soils were in fact taken into cultivation by this process, mere increase in arable area cannot have been the primary purpose of the introduction of plaggen manuring. The plaggen soils on damp ground and the extremely poor high sands are of later date than those on the moderately fertile dry sands (in particular the Brown Podzolic soils) - in other words, those soils which had already been under cultivation for long periods of time (Pape 1970; van de Westeringh this volume; Heidinga 1984a).

If plaggen manuring was practised in order to obtain a maximum return on animal manure, it must have been done to increase yield per agricultural unit. The other alternative is that the ground may have been exhausted and extra manuring was required to restore fertility to the original level. In this case, soil exhaustion must have manifested itself across the whole extent of the N German and Dutch sandy soils within the space of a century or two, which is only conceivable if external factors are enrolled as the cause. The agricultural land which in the Middle Ages was enriched by plaggen manuring had in part been in use since the Iron Age, while other areas were only cleared in the course of the Middle Ages, thus making a 'time bomb' explanation (predictable exhaustion in X number of years) rather less likely.

If, on the contrary, plaggen manuring was practised in order to improve the moisture balance of the soil, it would rather suggest an attempt to avoid harvest failure, i.e. the maintenance of regular production. In this case, the use of sods can only have been effective in the short term if fairly large quantities were spread at the outset. Furthermore, there is no reason to continue adding sods once a certain depth on the field has been achieved. The reasons for maintaining a particular system may, on the other hand, differ from those which led to the adoption of that system in the first place, quite apart from the possibility that a once successful method was simply continued without any particular awareness of its usefulness. If it is accepted that plaggen manuring was originally adopted in order to counter the effects of drought, then the rainfall shortage must have been of a serious nature. The odd dry summer can hardly have been an adequate reason to adopt such a costly system.

Whether it was climatic changes or else demographic and economic conditions which are to be linked to the introduction of plaggen manuring can only be adequately assessed if the date when the system first came into regular use can be established. Dating individual plaggen soils is not easy, since older, ploughed up archaeological relics are often incorporated in them and later material has also been transported downwards. The same is of course true for the organic materials which may be used for radiocarbon assay. Because of the background 'noise', especially of the earlier material, these soils often tend to be dated too early. According to Behre (1976), who has assessed the evidence from N Germany and the Netherlands, there is no support for any date prior to the 10th century. For Dunum, he was able to date the first plaggen manuring with a fair degree of certainty to the middle of the 10th century on a combination of archaeological, palynological and radiocarbon evidence. In Kootwijk too, there are indications for the adoption of this method of fertilization in this same period (see below). Since it would be far fetched to regard these two relatively well studied sites as chance exceptions, it does seem likely that a date in the 10th century for the inception of plaggen manuring on an extensive scale can be accepted.

The date in itself limits the number of possible causes for this agricultural revolution. Thus it is extremely unlikely that food shortages as a result of over population could have led to such drastic changes. There is no reason to suppose that the course of population density in the Netherlands and NW Germany differed in any marked degree from the general European trend, i.e. growth between ca. 600-850, stagnation between ca. 850-1050 and spectacular growth between ca. 1050-1300 (cf. Slicher van Bath 1963). There can be no question of a stress situation in the 10th century. Furthermore, until well into the 12th century there was ample vacant land open to clearance. Only from the 13th century are there hints that greater control was exercised over waste land on the sandy soils, as witnessed by stringent regulations concerning the use of this land. Neither can the production of surpluses for the non-agricultural population have been of significant importance prior to the 13th century. There were still scarcely any towns and even then, these were for the most part entirely dependent on the immediate neighbourhood for supplies. The considerable increase in feudal dues and other burdens, which reduced the agrarian population to a servile and hopeless condition in the 9th and 10th century would, according to Slicher van Bath (1963) in fact have removed any stimulus to increased effort and capital investment at this time. Measures intended to increase production could, of course, have been imposed on the peasantry from above. This is, for example, what is suggested in the case of Dunum, where plaggen manuring was introduced at the same time as a re-allotment of the settlement was carried out (Behre 1976). On the whole, however, the ownership of large estates was poorly developed in the areas where turf manuring found favour, so that such compulsory measures must have remained exceptional.

The only anomaly which I have been able to isolate in the 10th century which might have stimulated change in the agricultural system is a period of climatic change. Usually, it is damp and cold which is the cause of poor harvests in W Europe, but in this case, a period of abnormal aridity is concerned. Drought will cause major disruption to arable farming in areas which are by nature already dry. This was the situation at Kootwijk (prov. Gelderland, Netherlands) where the excavations carried out by the Albert Egges van Giffen Institute for Pre- and Protohistory of the University of Amsterdam have not only uncovered the consequences of these conditions, but have also provided evidence as to the nature of the climate at this time.

3.3 Drought and crisis at Kootwijk in the 10th century

Kootwijk lies in the central part of the Veluwe, a region which for the most part is formed of high, dry sands, approximately 1/5th of which is covered by sub-recent drift sands, which form what is probably the largest 'desert' of W Europe. Traces of settlement dating from the Neolithic to ca. 1150 AD have been found in the largest of the still-active drift sands of the region, the Kootwijkerzand, a little to the south of the village of the same name (fig. 3.1 a, b). Excavations carried out between 1971 and 1981 were largely concerned with the settlements from the Roman period onwards (see Heidinga 1984a). It is clear that from the Iron Age at least, arable farming was restricted to a small deliberately selected area, which is distinguished from the surrounding region by the presence of a relatively fertile soil (Brown Podzolic soils). The periodically shifting settlements which were the direct antecedents of modern Kootwijk were located on the edges of this fertile island. In the 8th century settlement was concentrated around a pool which was a vital source of water for man and beast. This settlement, which had grown to a village of more than 20 farms in the 9th century, disintegrated in the 10th century into smaller centres scattered elsewhere throughout the territory. Analysis has shown that barley, rye and oats were the major crops in the heyday of the settlement (Heidinga 1984a; Groenman-van Waateringe and van Wijngaarden-Bakker, forthcoming). Fertilization by means of turves was not yet practised. The 10th century saw several significant developments. In the first place, the pool began to dry out. Then, wind blown sand began to drift in over the fields, eventually, soon after 1000, frustrating every attempt at arable cultivation. Together, these factors were responsible for the flight of the villagers to other parts of the territory, where, however, safety was in the end only to be found on the site of the present day village of Kootwijk and in the low lying western sector of the area. Before the general abandonment, yet another change can be recorded: the formation of the first man-made soils. These last fields - though generally poor in pollen - have produced evidence solely for rye cultivation (Groenman-van Waateringe this volume). It is significant that all these developments can be linked to climatic extremes, the nature and extent of which became clear when the hydrology of the pool was analyzed.

The pool in question seems to have accumulated in a depression in the terrain where rain water collected on an impermeable bank ca. 75-150cm below the present surface. The 'true' water table occurs at a still greater depth. The water regime of the pool was, therefore entirely dependent on the balance between the rainfall on the one hand and evaporation, seepage and consumption on the other. The upper edges of the depression in which the rain water accumulated had always been dry, and it was in this part that wells were sunk as early as the 9th century in order to obtain clean drinking water. Every time that the water level receded, new wells would be sunk at a lower level, until eventually the centre of the by then entirely dried out pool was reached. Some of the wells could be dated archaeologically and taken with other evidence, in particular the different levels from which they were dug, a relatively detailed picture of the fluctuation of the water table could be formed.

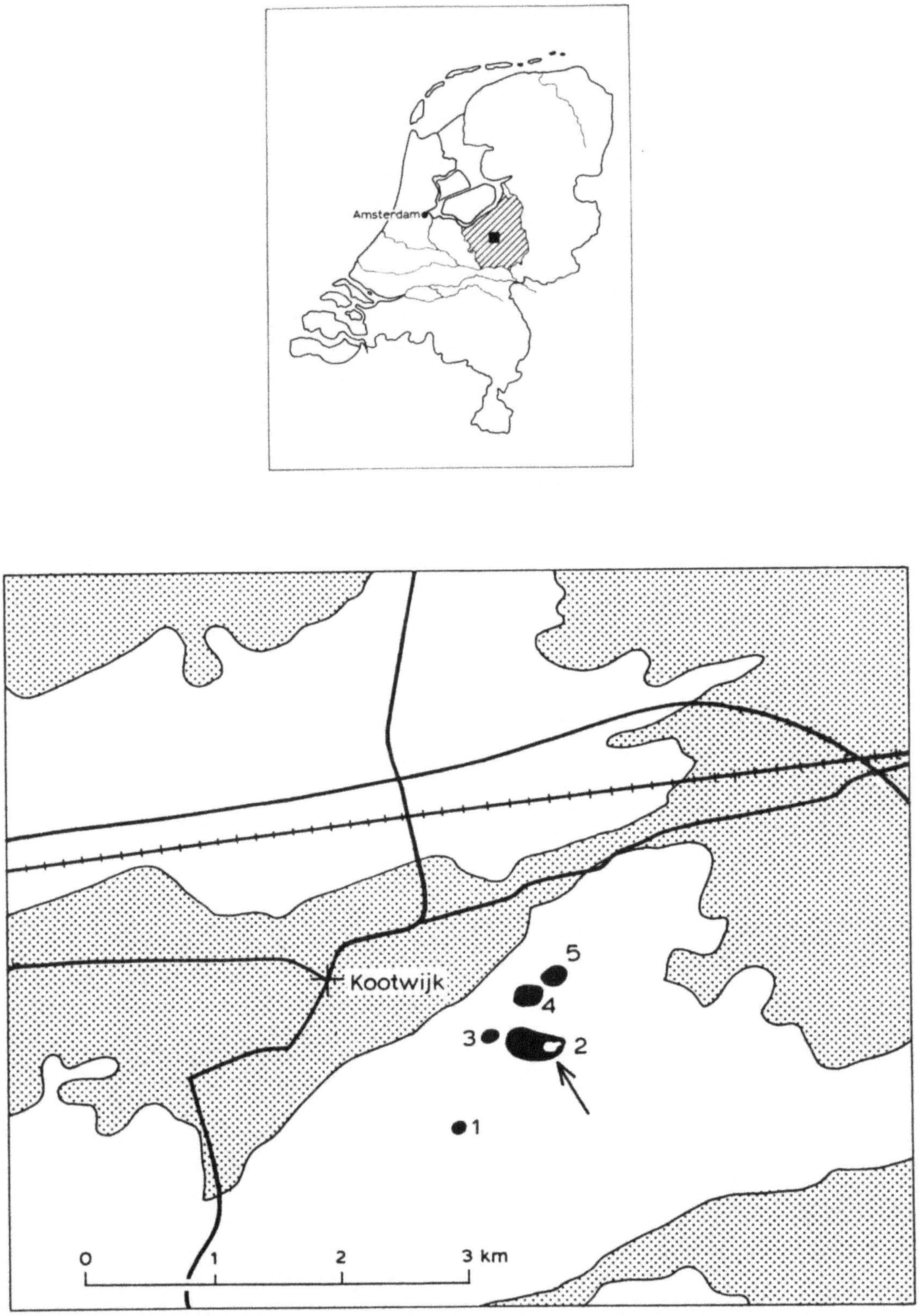

Fig. 3.1 The location of the Roman and medieval settlements revealed by excavation in the Kootwijkerzand 1: 11-12th century; 2: 8-10th century; 3: 2-3rd century; 4: 6-7th century; 5: 7th century. The location of the pool is indicated by the arrow. Drawing: IPP.

Plate 3.I The central area of the formal pool, with the latest well (a hollowed treetrunk) and arable land (the dark layer) from the 10th century AD, sealed by blown sand. Photo: IPP.

As it was possible to estimate the amount of water consumed and it could also be shown that seepage from the pool sides was negligible, the variable responsible for the ultimate drying out of the pool could be identified by a process of elimination. The cause can be attributed to climatic factors - through either decreasing rainfall and/or increased evaporation. Figure 3.2 charts the calculated annual rainfall between ca. 700 and 1000 (curve A: the rate of present climatic conditions is accepted; curve B: the deviation from the present rates is attributed for 50% to an increase in the rate of evaporation). The true values for rainfall would probably have lain somewhere between the two curves. It is likely that any shortage in rainfall will also be accompanied by increased evaporation, particularly in summer.

The conclusion must be that during some of the 10th century, and possibly even lasting into the 11th century, it was drier than has ever been the case since reliable weather records have been kept for any period of time in NW Europe. The climate over a large area of Europe must have had a continental character with cold dry winters and warm dry summers. Lamb (1982) reaches a similar conclusion by way of various methods of research, including the year ring frequency of the Spessart oaks, which display extreme stagnation in growth between c. 910 and 930 and again after 990. The consequences for arable farming on the high sands, where the risk of crop failure due to drought, even in normal circumstances, was already great, is easy to predict: poor yields of rye and a complete disaster as far as the summer cereals were concerned. The problem must have been compounded by the lasting character of the drought. Logically it might be assumed that a solution would be sought in a structural adaptation strategy. In this context, the marked dominance of rye which was noted at Dunum and - if less obviously, at Kootwijk - from the middle of the 10th century is significant. Rye is able to withstand periods of drought and also, as it is an autumn sown crop, it forms a protective cover for fields at risk of being blown out. The risk of wind erosion is greatest during dry spring weather when the fields are fallow or have just been sown with summer cereals. A cover of young winter cereals diminishes this particular hazard (see Steckhan 1950). Wind erosion can be further limited by reducing the period of fallow. Rye cultivation and intensification are, therefore, obvious methods of combating drought and the threat of wind erosion. The same is true of sod fertilization, which, indeed, increases the soil's moisture retention capacity. The period in which this method of fertilization was first systematically utilized was one of exceptional aridity. Conversely, no other factors emerge which might have stimulated such far reaching adaptations of existing farming practices and it is, consequently, tempting to think in terms of a causual link. Recurrent experience of drought perhaps led to the realization that a thicker plough soil diminished the likelihood of crop failure. This in turn may have led to the idea of actually increasing the thickness of the humic cover by means of soil brought in from elsewhere. This was of course most efficient when combined with the spreading of dung. The reason for the continuance of the practice for the next 900 years, however, lies in the gradual appreciation of other advantages of the system.

In Kootwijk things happened slightly differently. The plough soil which I regard as a man-made soil was raised on a subsoil of blown

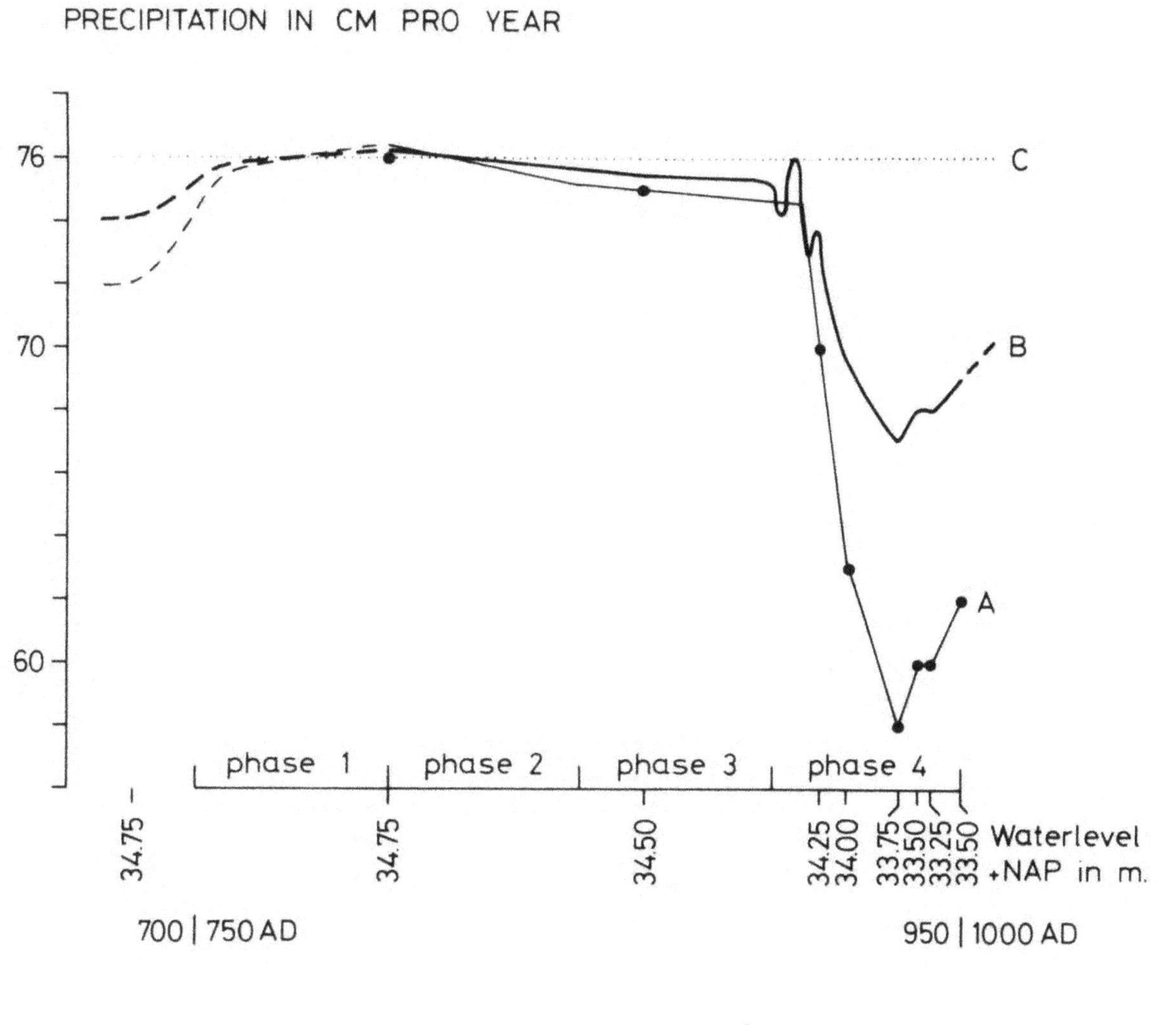

Fig. 3.2 Estimated annual precipitation between ca. 700 and 1000 AD. A: minimum (evaporation rate as at present); B: maximum (symmetrically increased evaporation); C: present day quantity. Drawing: IPP.

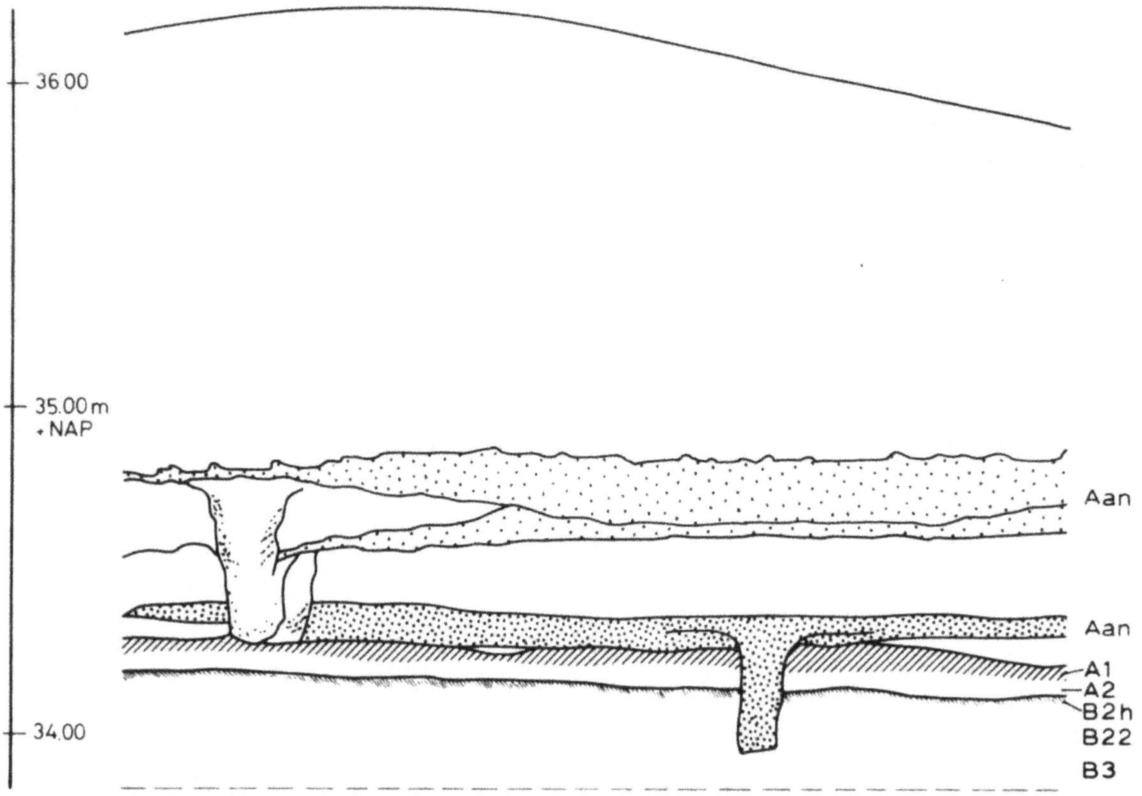

Fig. 3.3 Section through man-made agricultural layers (Aan) from the 10th century AD in the Kootwijkerzand. Both layers lay on and are blown over by drift sand. Drawing: IPP.

sand. Presumably this is primarily an attempt to restore the earlier conditions. Figure 3.3 shows a humic plough soil of the normal thickness (10-15cm) raised on top of a layer of blown sand which can only have been deposited shortly before. This arable layer was in turn covered by drifting sand. On this sand too, a new, moderately humic plough soil was laid which must originally have attained a depth of some 40cm (the top has been blown out in many places). This depth is partially explicable in terms of periodic sand drifting, but, in my opinion, it also points to human agency in an attempt to collect as much humic soil as possible on the field. The artificial nature of this plough soil is called into question by Groenman-van Waateringe (this volume) and indeed many of the characteristic properties of a classic plaggen soil are absent. This may be due to the nature of the material used to raise the field. The second plough soil was certainly not composed of heather sods and the general impression is rather that humic soil from (abandoned) fields in the neighbourhood had been collected. It does, nevertheless, seem to me that it is highly unlikely that the process of soil formation could be so far advanced in the drift sands after only a couple of decades to account for either the humic nature of the lower soil or the thickness of the upper plough soil. Neither could they result from normal manuring practices. Selective natural deposition of the humic particles alone in this particular area seems inherently unlikely.

3.4 Conclusion and some final comments

However dubious the status of the 10th century plough soil beneath the drift sands at Kootwijk as a genuine plaggen soil may be, it is at any rate certain that this method of soil enrichment began to be used elsewhere in this period. That climatological conditions stimulated its adoption on a large scale cannot be proved, but, considering the evidence available for the 10th century, it is a possibility which is certainly worth consideration. The explanatory model proposed also accounts for the relationship between sod fertilization, rye cultivation and intensification. One question remains unsolved: why were plaggen soils not favoured in other drought sensitive areas of Europe which were subject to the same climatological conditions? Although an answer is not easy to find, the significance of another characteristic feature of medieval arable cultivation should not be underrated in this context. This is the ridge-and-furrow which can still be identified over great tracts of Western and Central Europe. The special way of ploughing results in the accumulation of humic soil in long, narrow beds, domed in cross section. Such beds possess better moisture retention properties than a flat ploughed field. That ridge-and-furrow and plaggen soils should be regarded as variants of one and the same system is something which at this stage of research I would offer merely as a suggestion.

Note

This article is a summary of Heidinga 1984a III, 2 and 3. See also Heidinga 1984b.

Acknowledgements

Thanks are due to Drs. Carol van Driel-Murrey for translating the text into English, André Numan for the drawings, Fred Gijbels for the photograph, and Professor Willy Groenman-van Waateringe for the discussions on this subject.

References

Behre, K.E., 1976. Beginn und Form der Plaggenwirtschaft in Nordwestdeutschland nach pollenanalytischen Untersuchungen in Ostfriesland. Neue Ausgrabungen und Forschungen in Niedersachsen 10, 197-224.

Boserup, E., 1965. The conditions of agricultural growth. London: Allen and Unwin.

Groenman-van Waateringe, W. and Wijngaarden-Bakker, L.H. van, (eds.), in press. The ecology of early medieval Kootwijk. Cingula.

Heidinga, H.A., 1984a. De Veluwe in de vroege middeleeuwen. Aspecten van de nederzettingsarcheologie van Kootwijk en zijn buren. Ph.D. thesis, Univ. of Amsterdam. (An edition in English is forthcoming).

Heidinga, H.A., 1984b. Indications of severe drought during the 10th century AD from an inland dune area in the Central Netherlands. Geologie en Mijnbouw 63, 241-248.

Kossack, G., Harck, O. and Reichstein, J., 1974. Zehn Jahre Siedlungsforschung in Archsum auf Sylt. Bericht der Römisch-Germanischen Kommission 55, 261-428.

Lamb, H.H., 1982. Climate, history and the modern world. London and New York: Methuen.

Pape, J.C., 1970. Plaggen soils in the Netherlands. Geoderma 4, 229-255.

Slicher van Bath, B.H., 1963. The agrarian history of Western Europe AD 500-1850. London: Arnold.

Steckhan, H., 1950. Bodenabtrag durch Wind in Niedersachsen - eine Gefahr für die Landeskultur. Neues Archiv für Niedersachsen 17, 313-335.

Tiesing, H., 1974. Over landbouw en volksleven in Drenthe. 2 vol. Ed. by C.H. Edelman. 2nd ed. Assen: van Gorcum.

Westeringh, W. van de, 1978-1979. De ontwikkeling van een agrarisch cultuurlandschap in een zandgebied (Winterswijk). Cultuurtechnisch Tijdschrift 18, 199-210.

4. POLLEN FROM PLAGGEN SOILS IN THE PROVINCE OF NORTH BRABANT, THE NETHERLANDS

C.C. Bakels

Abstract

Plaggen soils are man-made arable soils. They are characterized by thick A horizons, which owe their thickness to an age-long practice of using manure containing a fair amount of mineral matter. Plaggen soils abound in pollen, but the pollen spectra are difficult to interpret. The reason for this is that the pollen sources are interwoven to a high degree. It is, however, possible to trace some of the origins of the matter of which the manure is composed and to distinguish some of the crops grown. The dominance of rye cultivation during the Middle Ages and later historic periods stands out clearly.

4.1 Introduction

Plaggen soils are arable soils with an unusually thick A horizon (Pape 1970). They are man-made. The thick A horizon is the result of a special way of preparing manure.

To obtain as much manure as possible, the farmer keeps his animals on a bedding composed mainly of turves cut from either heath or grassland. To this clean sand, fallen leaves collected from the forest and dry ditches, and even straw - if not too expensive - may be added. The farm animals drench this matter with their droppings. They are kept in the stables for as long as possible, at least during the night. The mixture of bedding and droppings can become quite thick in the course of time. Therefore, the floor of the stables is deepened with respect to the floor of the living quarters. Once or twice a year the manure is removed. Later on loamy earth or more turves may be mixed in to increase the amount of matter.

This way of preparing manure is best known from the sandy soils in NW Europe. The product was used for fertilizing poor arable land. The practice was common from the Middle Ages up till the beginning of the 20th century. Because the manure contains a fair amount of mineral matter, which will never decay, the surface of the fields was raised. Nowadays these fields form low mounds in the landscape. The soils in question are called plaggen soils after the German word for turves: *Plaggen*. The colour of these soils varies from black to brown.

Plaggen soils abound in pollen. Contrary to what one would expect to encounter in dry sandy soils, which have, moreover, been ploughed repeatedly, the pollen grains are beautifully preserved. The reason

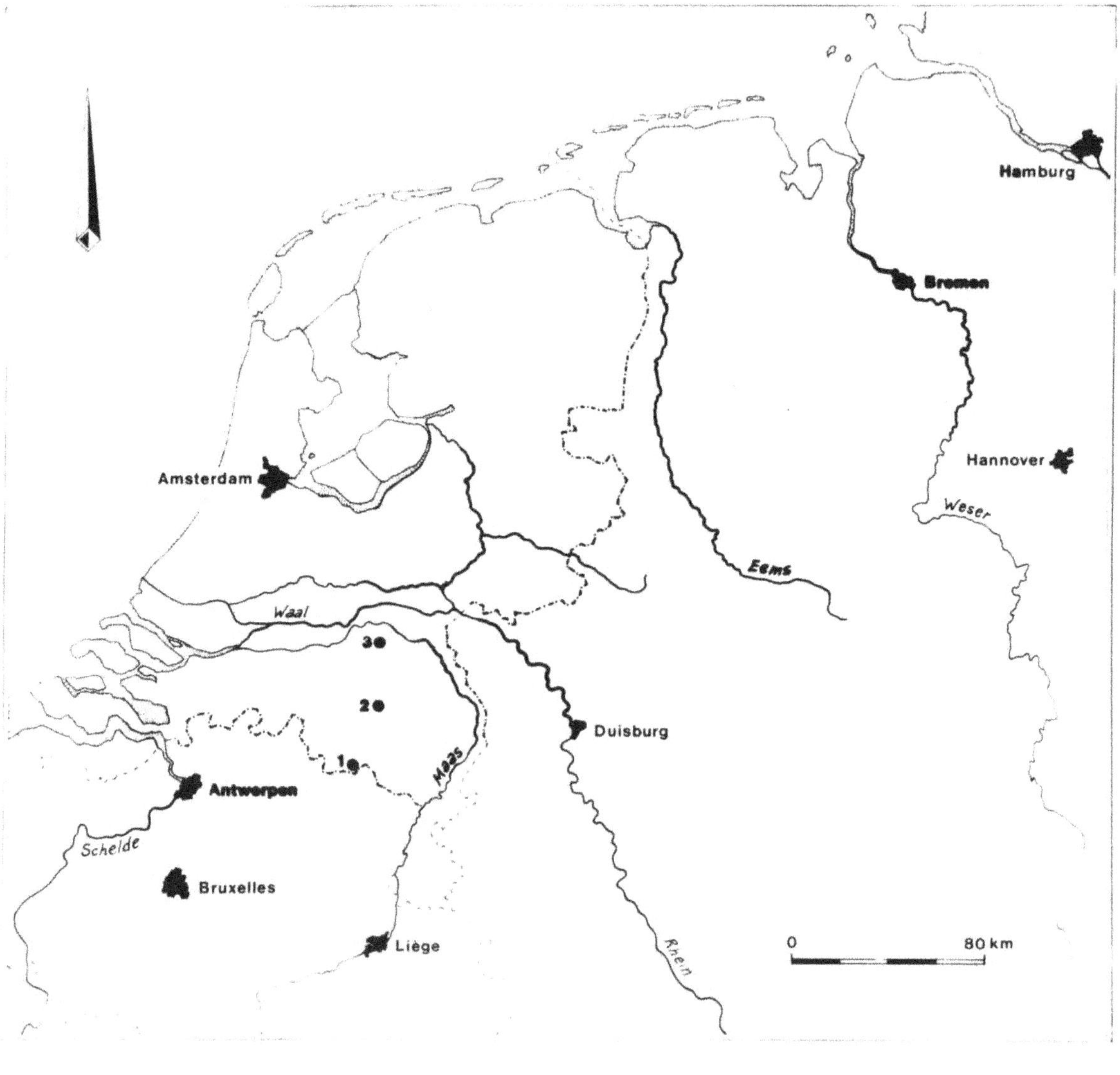

Fig. 4.1 The plaggen soil area in northwestern Europe (shaded) and the location of the sites mentioned in the text. Drawing: IPL. 1. Borkel en Schaft. 2. Son en Breugel. 3. Oss.

why is not quite clear but one point may be that the pH is less than 5. The pH from the plaggen soils investigated for this paper ranges from 3.6 to 4.3.

Out of curiosity I investigated the pollen content of three plaggen soils in the Dutch province of North Brabant. Their location within the NW European plaggen soil area is indicated in fig 4.1. The samples were collected from the walls of trenches.

4.2 The pollen diagrams

The slides were prepared as usual when handling mineral matter. The only difficulty encountered while drawing the diagrams was selecting a pollen sum. There is no clear difference between local and regional pollen. The pollen in the plaggen soil has at least three main sources. The first is the usual pollen rain which originates in the near or more distant surroundings. The second is the crop grown on the field together with its field weeds. The third source is the manure and this is a most complicated source. It comprises pollen from the bedding, i.e. turves, fallen leaves, straw, but it also contains pollen from the droppings, i.e. from the food consumed by the stabled animals. Some of this is illustrated in fig. 4.2. The actual vegetation contributes to the pollen record in more than one way. It is therefore impossible to select a pollen sum, with the exception of the total pollen sum. The three diagrams which follow were drawn on the basis of this sum.

4.2.1 Borkel en Schaft

The plaggen soil from Borkel en Schaft is a so-called black plaggen soil. It shows two horizons: an Aanp (anthropogenic A, ploughed), which is the recent and subrecent ploughed horizon, and an Aan2 beneath. The plaggen soil rests on a xero-podzol that developed in coversand. Samples were taken from both the original soil profile and the anthropogenic soil. The diagram is shown in fig. 4.3. Species not mentioned are to be found in Table 4.1.

The original C and BC horizons do not contain pollen, but the B and A horizons do. They are characterized by high values for Ericales (mainly *Calluna*), Gramineae, *Rumex acetosella* and some *Potentilla*. This assemblage must have belonged to the original vegetation, which seems to have been a heath. This is in agreement with the presence of a xero-podzol. The high values of *Rumex acetosella* imply that this heath was disturbed to some extent. Trees play no role, even though their pollen percentages are the highest of the whole diagram here. The country was open, at least in the area of sampling. It is, of course, very difficult to assess the extent of forest in more distant areas from this type of sample.

The B horizon already shows pollen types which belong to cultivated species. These could have made their way down from the first fields above in a natural way, but this is not the only possibility. The trenches showed neat rows of spade-marks at the base of the plaggen soil. The pollen grains may have been dug down,

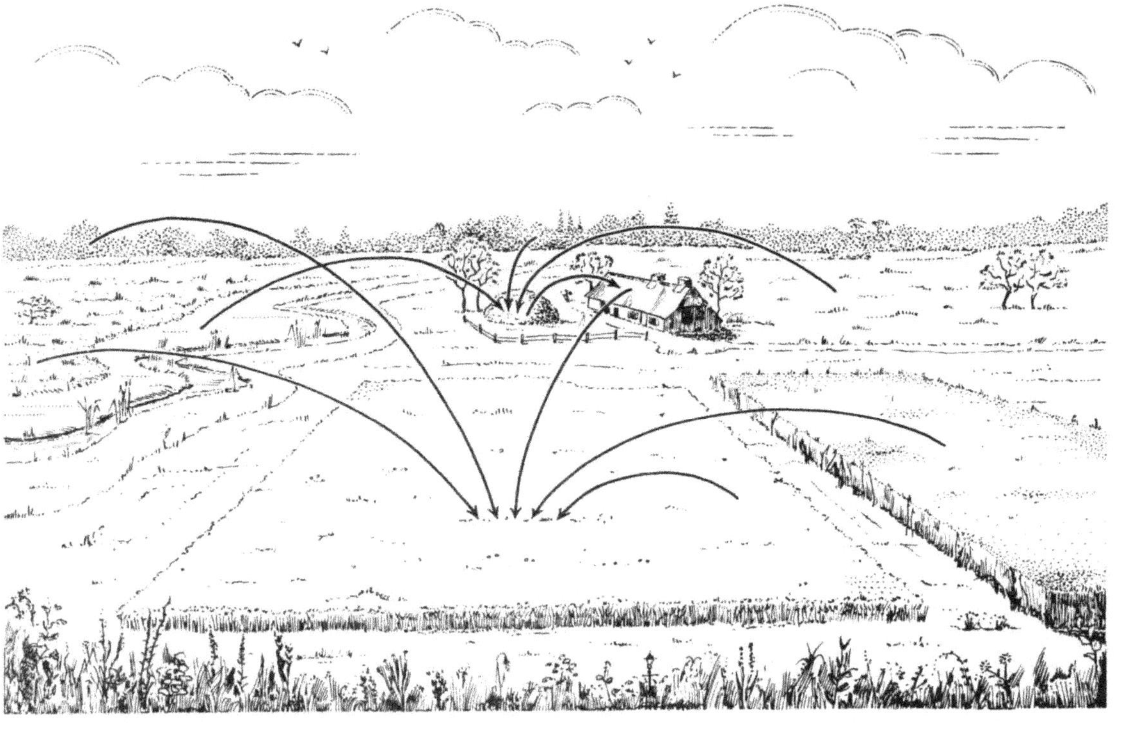

Fig. 4.2 Some of the possible ways in which pollen may find its way to the arable land. A heap of turves and leaves next to the farm house awaits use as bedding in the stable. Not shown are the animals adding matter by means of their droppings. The droppings contain pollen taken in with the food browsed outside or eaten with the fodder distributed in the stable. The resulting manure is carted onto the land where it adds pollen to the local and regional pollen rain. Drawing: IPL.

although these spade-marks were avoided during sampling. Anyhow, they represent the first crops documented by pollen: Hordeum (listed under Cerealia) with some Secale.

The Aan horizon itself must have been disturbed over and over again by ploughing. Nevertheless, the pollen curves are not merely straight lines. There is a striking difference between the Aan2 and Aanp horizons, which can be explained by assuming that some of the pollen in the Aanp horizon was ploughed into the soil whilst almost no new mineral matter was added. The pollen content of this layer is therefore a mixture of pollen which is the result of plaggen manuring and pollen from another, later method of fertilization without or almost without this kind of manure. From this point of view the Aan2 zone represents the original pollen assemblage in a plaggen soil better.

One of the main sources of turves must have been the heath. Ericales (Calluna with a little Erica) amount to no less than 30%. A second source might have been grassland, but the Gramineae pollen could also have come from Molinia growing on the heath. The third plant with a rather high percentage is Rumex acetosella. This sheep sorrel might have grown on disturbed patches on the heath, but another possibility is that it was a field weed.

A second category of bedding is represented by the Monoletae spores. Their origin lies either in the use of matter collected in forests or dry ditches, or in the use of fern leaves. The percentage is rather low, and so are the percentages of tree and shrub pollen which we would expect to find if fallen leaves were indeed used. The source was perhaps not very important, or, another possibility, the pollen content of the leaf-mould was meagre. According to experiments performed by Havinga (1984), pollen in leaf-mould decays within twelve months.

A third category of bedding which turns up in the manure is Sphagnum peat. It is known that dried, crumbled peat was also brought into the stables.

The crop plants are represented by rye (Secale), other Cerealia (barley with a few oats), buckwheat (Fagopyrum) and spurrey (Spergula arvensis). Spurrey was an important crop plant. It was sown as fodder and had the reputation of being especially good for dairy cattle. Of course it is also a field weed.

In addition to the plants already mentioned, the Aanp also shows maize (Zea) and hemp (Cannabis). There is a clear increase in buckwheat and in two other pollen types which might represent crops: Cruciferae (rape-seed and the like) and Chenopodiaceae (beets). These two may have been weeds, but in that case the Chenopodiaceae would suggest, on ecological grounds, the cultivation of root crops.

Quite interesting is the behaviour of the cornflower (Centaurea cyanus). It is often stated that cornflower is always found together with rye. Here this is apparently not the case. It comes later in the pollen record.

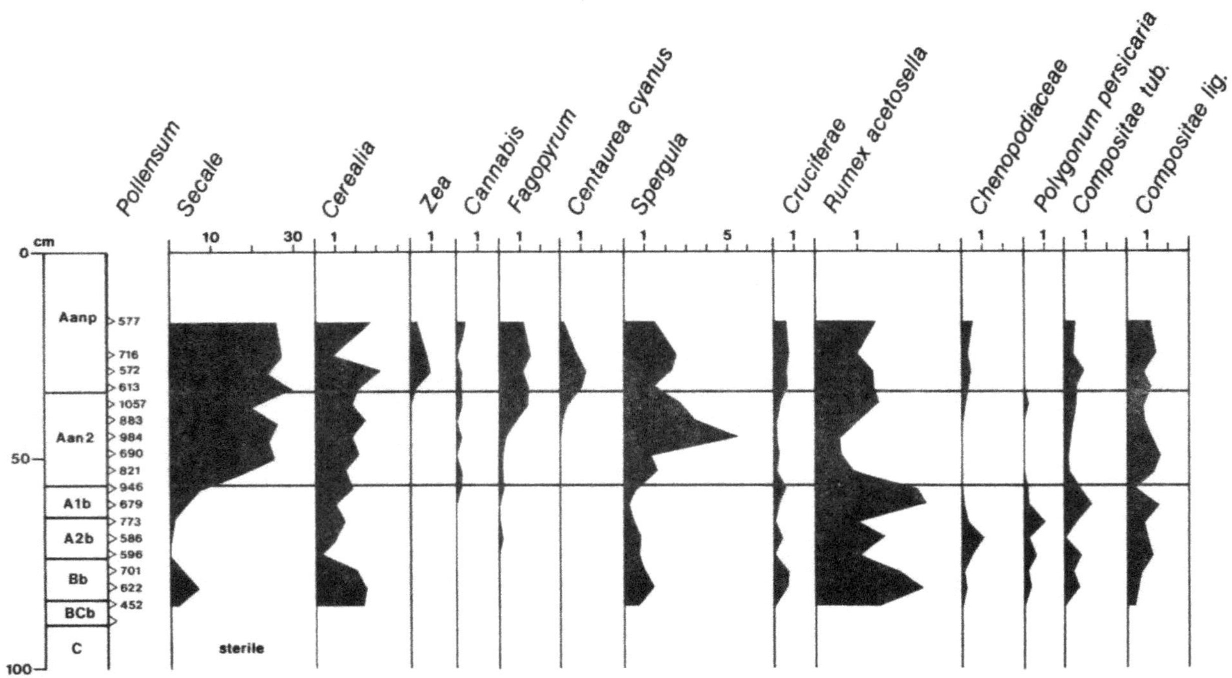

Fig. 4.3a The pollen diagram from Borkel en Schaft. Drawing: IPL.

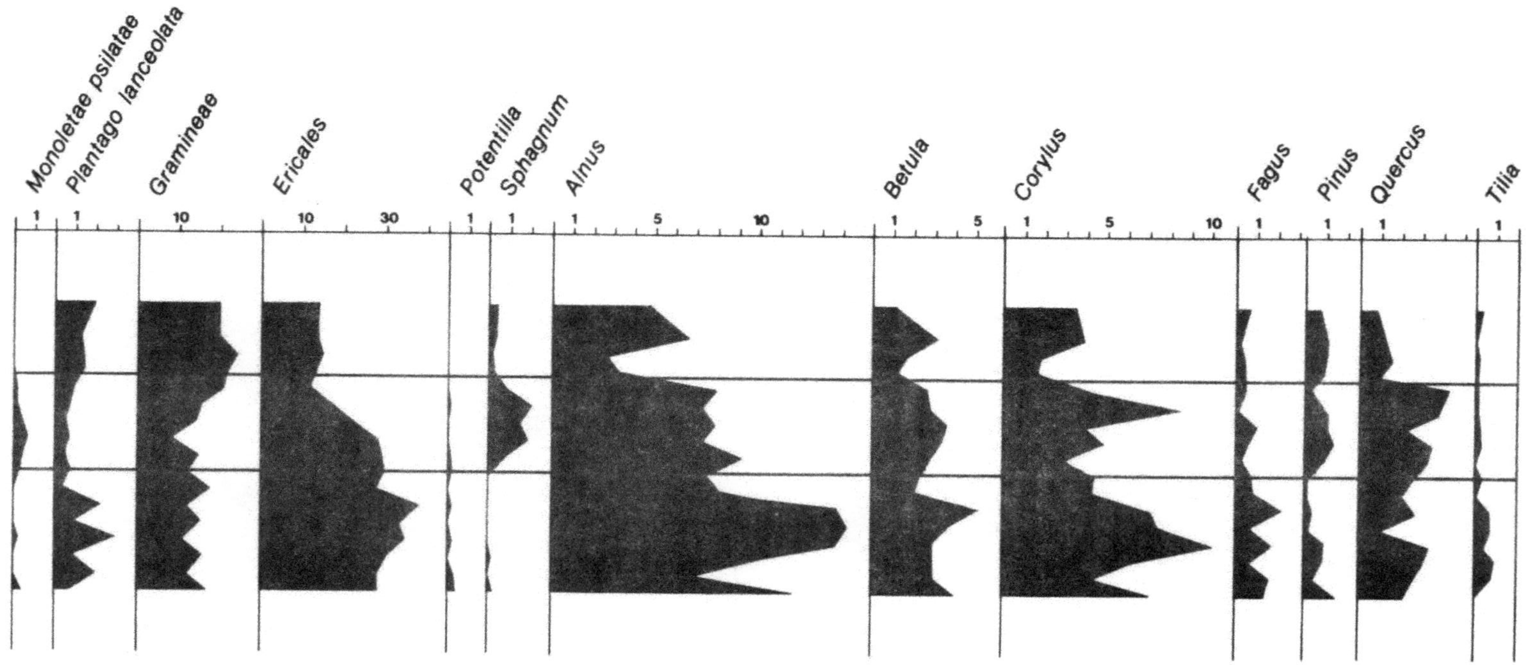

Fig. 4.3b The pollen diagram from Borkel en Schaft continued. Pollen types mainly after Faegri and Iversen 1964. Drawing: IPL.

	Borkel en Schaft					Son en Breugel					Oss			
	B	A2	A1	Aan2	Aanp	B	Aan4	Aan3	Aan2	Aanp	B	Aan3	Aan2	Aanp
<u>Trees and shrubs</u>														
Carpinus	+	+	+	+	+	.	.	.	.	+	.	+	+	+
Fagus	+	+	+	+	+	.	+	+	+	+	.	+	+	+
Fraxinus	+	+	+	+	+	.	.	.	.	.	.	.	.	.
Hedera	+	.	.	.	.	.	.	.	.	.	.	.	.	.
Ilex	.	.	.	.	.	+	+	+	+	+	.	.	.	.
Juglans	.	.	.	+	.	.	.	.	.	.	.	.	.	.
Ligustrum	.	.	.	+	.	.	.	.	.	.	.	.	.	.
Myrica	.	+	.	+	.	.	.	.	.	.	.	.	.	.
Picea	+	+	.	+	+	.	.	.	.	.	.	.	.	+
Prunus	+	.	.	+	+	.	.	.	.	.	.	.	+	+
Sambucus nigra	.	.	.	.	.	.	.	.	.	.	.	.	.	+
Salix	.	.	+	+	+	.	+	+	+	+	.	+	.	+
Tilia	+	+	+	+	+	+	+	+	+	+	.	.	.	+
Ulmus	+	+	+	+	+	.	+	+	+	.	.	+	.	+
<u>Herbs</u>														
Artemisia	+	+	+	+	.	+	.	.	+	+	.	.	.	+
Campanula	.	.	.	+	.	.	.	.	.	.	.	.	.	.
Cannabis	.	.	+	+	+	.	.	.	.	.	.	.	.	.
Caryophyllaceae	+	+	+	+	+	+	+	+	+	+	+	+	+	+
Centaurea jacea	.	.	.	+	+	.	.	.	.	.	.	.	+	+
Cerealia (*	+	+	+	+	+	+	+	+	+	+	+	+	+	+
Cyperaceae	.	+	+	+	+	+	.	.	.	+	.	+	.	.
Epilobium	.	.	.	.	.	.	.	.	.	+	.	.	.	+
Filipendula	+	+	+	+	+	.	.	.	.	.	.	.	.	.
Genista-type	.	.	+	.	.	.	.	.	.	.	.	.	.	.
Gentiana pneumonanthe	+	.	.	.	.	.	.	.	.	.	.	.	.	.
Geraniaceae	.	.	.	.	.	+	+	+	+	.	.	+	+	+
Jasione	+	.	.	.	.	.	.	.	.	.	.	.	.	.
Linum	.	.	.	.	.	.	.	.	.	.	.	.	.	+

Table 4.1 Pollen types not mentioned in the pollen diagrams. Types mainly after Faegri and Iversen 1964.

	Borkel en Schaft					Son en Breugel					Oss			
	B	A2	A1	Aan2	Aanp	B	Aan4	Aan3	Aan2	Aanp	B	Aan3	Aan2	Aanp
Lotus corniculatus	.	.	.	.	.	.	.	.	.	+	.	.	.	.
Lysimachia-type	.	.	.	.	+	.	.	.	.	.	.	.	.	.
Mentha-type	.	+	+	.	.	.	.	+	.	.	.	.	.	.
Menyanthes	.	.	.	.	+	.	.	.	.	.	.	.	.	.
Onobrychis-type	.	.	.	.	.	.	.	.	+	.	.	.	.	.
Papaver	+	+	+	+	+	+	.	.	.	.	.	.	.	.
Plantago media	.	.	+	+	+	.	+	.	.	.	.	.	.	.
Polygonum aviculare	+	.	.	.	+	+	+	+	+	+	.	.	.	+
Polygonum convolvulus	.	.	+	+	+	.	.	.	+	+	.	+	+	+
Polygonum persicaria-tp	+	+	+	+	.	+	+	+	.	+	+	+	+	+
Potentilla	+	+	+	+	.	+	+	+	.	+	.	.	.	+
Prunella-type	.	.	.	+	.	.	.	.	.	.	.	.	.	.
Radiola	.	.	.	.	.	.	+	.	.	+	.	.	+	.
Ranunculus	+	+	.	+	.	+	+	+	+	+	+	+	+	+
Rubiaceae	+	.	+	+	.	+	+	.	.	+	+	+	+	+
Rubus	.	.	+	+	+	.	+	.	.	.	.	.	.	+
Rumex	.	+	+	.	+	.	.	.	.	.	.	.	.	.
Spergularia	+	.	.	.	+	.	+	+	+	.	.	.	.	.
Stachys-type	.	.	.	+	.	.	.	.	.	.	.	+	.	.
Succisa	+	+	.	+	.	.	.	.	.	.	.	.	.	.
Thalictrum	.	.	.	.	.	.	.	.	.	.	.	.	.	+
Trifolium-type	+	.	.	+	+	+	+	+	+	+	+	+	+	+
Typha latifolia	.	.	.	.	.	.	+	.	.	.	.	.	.	.
Umbelliferae	.	.	.	+	+	.	+	+	.	+	.	.	.	+
Urtica	+	+	+	+	+	.	.	.	.	.	.	.	.	.
Vicia-type	.	.	.	.	+	.	.	.	.	+	.	+	+	+
Viola arvensis	.	.	.	.	.	.	.	.	+	+	.	.	+	+
Zea	.	.	.	.	+	.	.	.	.	+	.	.	.	.
<u>Ferns, club-mosses and scale-mosses</u>														
Anthoceros laevis	.	.	.	.	+	+	+	+	+	.	.	+	+	+

	Borkel en Schaft					Son en Breugel					Oss			
	B	A2	A1	Aan2	Aanp	B	Aan4	Aan3	Aan2	Aanp	B	Aan3	Aan2	Aanp
Lycopodium complanatum	.	.	.	.	.	.	.	+	.	+	.	.	.	.
Monoletae psilatae	+	+	+	+	.	+	+	+	+	+	+	.	+	+
Polypodium	.	.	.	.	.	+	+	+	+	+	.	.	+	+
Riccia	.	.	.	.	.	+	+	+	+	+	+	.	+	+

(* Secale and Zea excluded)

Of all possible pollen sources which may have contributed to the total pollen assemblage, the pollen rain itself remains obscure. Its effect is masked by the other ones and even the boundary between the sources mentioned above is not as clear as suggested. The animal droppings must have played an important part. If cattle was fed on, for instance, rye straw, the origin of some of the rye pollen will lie in the droppings and not in the crop on the actual field sampled. Likewise, the Calluna pollen in the manure not only originated in the turves, but also in sheep droppings. The picture as a whole cannot be unravelled as much as we would like.

4.2.2 Son en Breugel

The plaggen soil of the "Hooidonksche Akkers" near Son en Breugel is a so-called brown one. The original soil beneath is a moder podzol in coversand. As in Borkel en Schaft, the buried C and BC horizons are sterile. The B horizon contains pollen. The original A horizon has vanished. Thin sections prepared by A. Jongerius at Wageningen show that the B horizon got mixed with material from above, so that at least part of its pollen comes from the Aan4 horizon. Nevertheless, several types belong exclusively to the B horizon and are evidence of a former vegetation (fig. 4.4). These are, to begin with, Polypodium and the Monoletae psilatae. They could be the last remnants of a local forest. Havinga (1984) has shown that Polypodium is very resistant to decay in soils with high biological activity. The forest may have been succeeded by a weed vegetation from which Plantago lanceolata, Scleranthus and Polygonum aviculare have survived, but these three may also have grown in the first fields. Most of the Secale must have been mixed in with the matter introduced from above.

The man-made arable soil reveals four layers. These were distinguished in the field on the basis of slight differences in colour. The Aan4 is mixed with the A horizon of the original profile. This was apparent both in the field and in the thin sections. The Aan3 horizon is more homogenous. According to the micromorphology, this part of the Aan horizon has a large component of river loam. The Aan2 horizon is coarser in structure. All three layers contain fragments of peaty material. The boundaries between the layers are not sharp. The Aanp horizon on the contrary, stands out clearly. It is very homogenous. Its matter is pulverized as a result of ploughing.

The Aan4, Aan3 and part of the Aan2 layers do not show much difference in their pollen content. The most obvious changes are shown by the curves of Ericales and Cruciferae. The Ericales (Calluna and Erica) rise in the Aan4 and fall in the Aan3 and Aan2 horizons. The decline is perhaps the result of a more intensive use of grass turves cut in the valley of the nearby rivulet. The loam that adhered to the grass roots could be the source of the loam found in, at least, the Aan3 horizon. Anyhow, heath and heath turves played a less important role in the Son en Breugel agriculture than in that of Borkel en Schaft. The percentages of Ericales are much lower.

A second component of the bedding in the stables was peat. Its use is demonstrated by both the Sphagnum spores and the fragments of peat discovered in the thin sections.

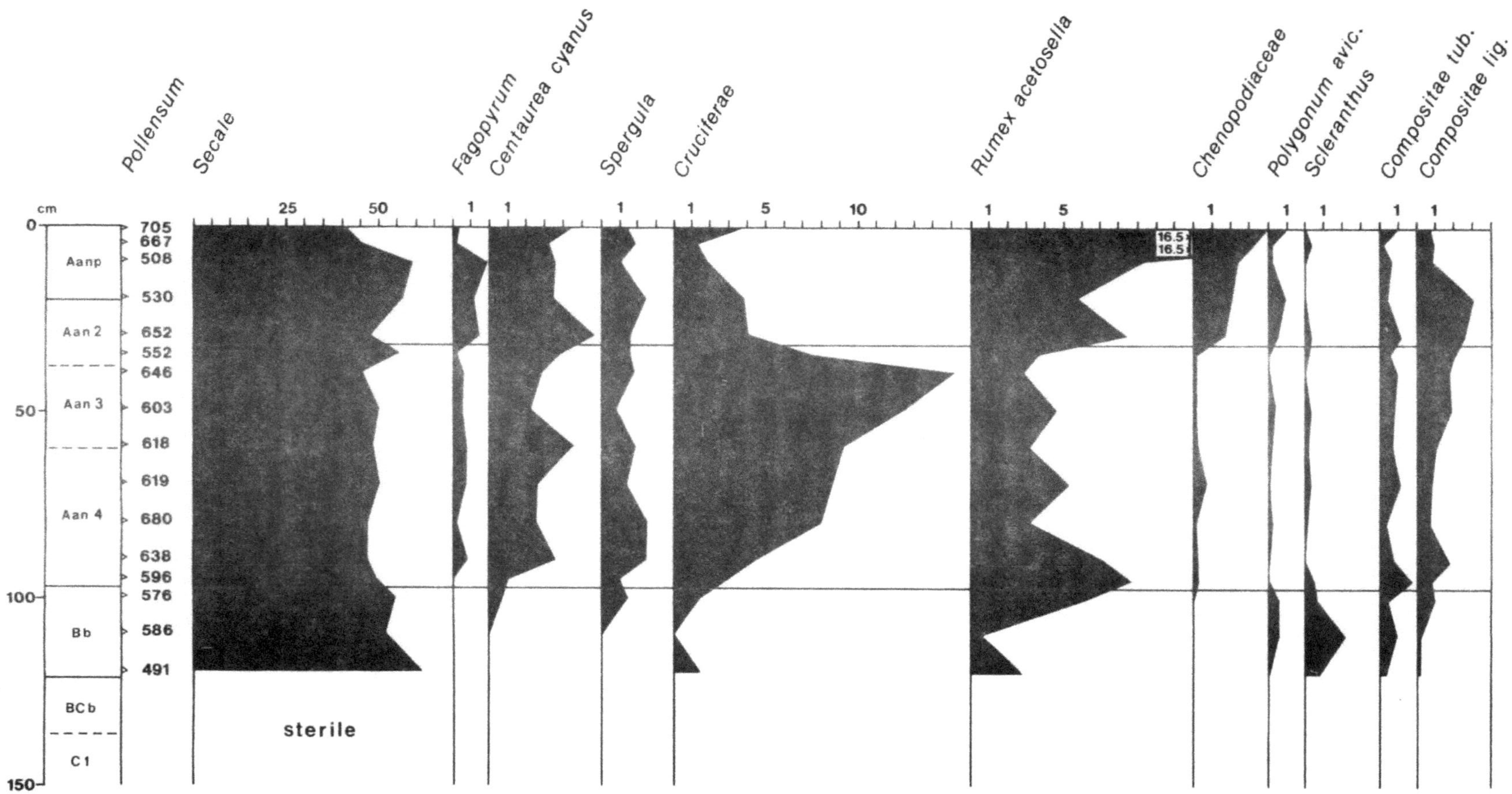

Fig. 4.4a The pollen diagram from Son en Breugel. Drawing: IPL.

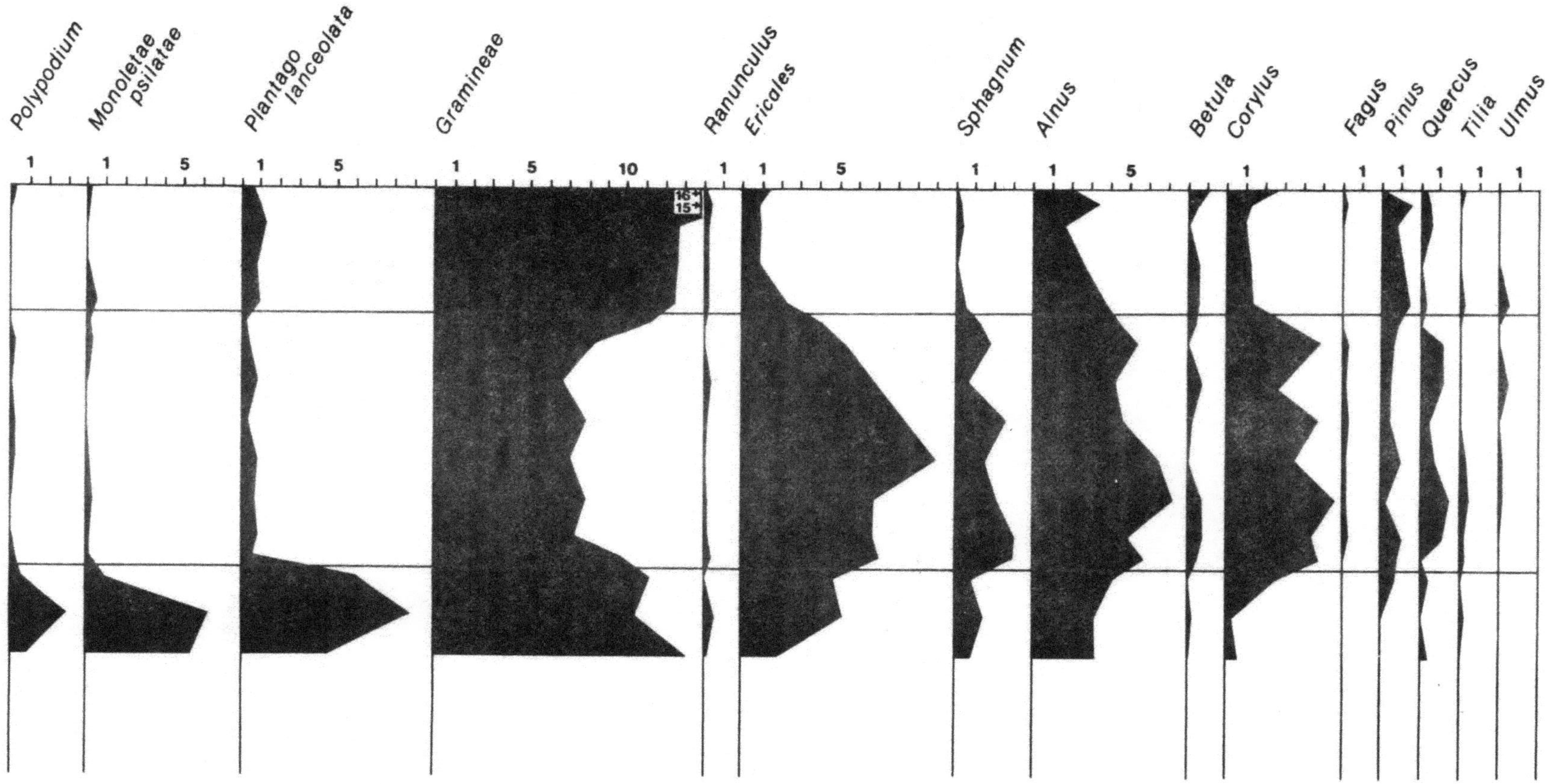

Fig. 4.4b The pollen diagram from Son en Breugel continued. Drawing: IPL.

The increase in Cruciferae pollen towards the end of the accumulation phase Aan4 to Aan2 is perhaps due to a growing importance of crop plants belonging to this family. Other crop plants are rye (Secale), dominant as usual, spurrey (Spergula), buckwheat (Fagopyrum) and some barley and oats. It would appear that buckwheat and possibly spurrey as well were not grown from the very beginning. Their curves suggest a younger history for these crops in comparison with rye. Centaurea cyanus appears later too.

As in Borkel en Schaft, the uppermost horizon, Aanp, is characterized by different pollen spectra. The upper part of the Aan2 horizon closely resembles the Aanp layer. This part is thought to belong to the Aanp horizon rather than to the actual Aan2 layer described in the field. The overall thickness is 38cm which comes fairly close to the thickness of the Aanp horizon in Borkel en Schaft: 35cm. The curves suggest a greater importance of buckwheat and root crops (Chenopodiaceae) and a lesser use of Cruciferae. Grasses and sheep sorrel may have been real pests. The presence of maize pollen and clover (Trifolium) indicates the introduction of new crops.

An attempt has been made to date this plaggen soil by means of the sherds found in it. The work was carried out by J. Assendorp with the assistance of members of the amateur society AWN. They excavated part of the plaggen soil and collected all sherds in layers 10cm thick. The result is shown in fig. 4.5

The prehistoric sherds and those belonging to the Roman Period and the 8-9th century were distributed mainly in the lower part of the Aan4 horizon. This was to be expected because they must have lain on the original surface and this surface was included in the Aan4 horizon. Of course many of the sherds had been dispersed in the profile by ploughing.

The regular distribution of ceramics in the plaggen soil begins with material attributed to the 10-11th century. The Aan4, Aan3 and Aan2 horizons contain a mixture of sherds ranging from the 10th to the 15th century. The Aanp layer produced sherds from the 16th century onwards. The difference between this horizon and the lower ones is as clear as in the pollen diagram.

The sherds must have been disturbed by ploughing along with the pollen grains and can therefore be used to date the pollen zones. This means that the lower zone contains mainly pollen from a period before the 10th century. The rye in the B horizon may therefore represent a 9th century crop. It is less probable that it originates from a Roman or Iron Age crop, because this pollen must have lain at the surface or near the surface for a long time and would have decayed before the plaggen soil was formed.

The middle zone, with its dominant rye and its buckwheat, spurrey and Cruciferae, dates back to the Middle Ages. The upper zone is younger. There are marked differences between the middle and upper zone. The sherds suggest that the accumulation of soil came to an end. All the 16th century and later material may have been ploughed down from the surface. If this is true, this means that this particular patch of land was not fertilized with plaggen manure very much after

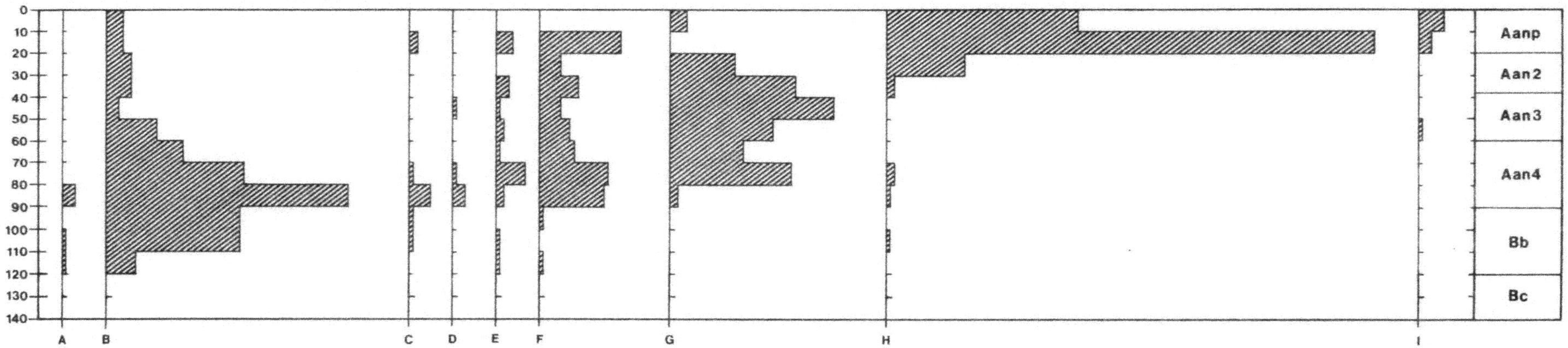

Fig. 4.5 Sherd diagram from the plaggen soil at Son en Breugel. The sherds are given in numbers for layers 10cm thick and per square metre. The lowest unit is one sherd. Drawing: IPL.

A = Neolithic + Bronze Age
B = Iron Age
C = Roman Period
D = 8-9th century
E = 10-11th century
F = 12-13th century
G = 14-15th century
H = 16-18th century
I = subrecent and recent

the 15th century, which is in contrast with the general trends in North Brabant. Up till now it has always been thought that the use of turves continued until the 20th century. Investigations making use of written sources may throw more light on this problem.

4.2.3 Oss-IJsselstraat

A third diagram was obtained from the plaggen soil in the IJsselstraat in the town of Oss. The buried soil profile is a hydro-podzol, which once again developed in coversand.

The C horizon was not sampled but in other samples taken on the same location this horizon was found to be sterile. The B horizon is present, but the A horizon has merged with the first plaggen soil. Here again the difference between an Aan3 and an Aan2 horizon is merely a question of a slight difference in colour. This layer is separated from the Aanp horizon by a sharp line. As in the other diagrams, the pollen zones correspond to the horizons outlined in the soil profile (fig. 4.6).

The buried B horizon is dominated by pollen of Gramineae and *Rumex acetosella*. The grasses may be the representatives of a former local grassland. Sheep sorrel is a pioneer of newly opened sandy soil and may reflect the first reclamation of the land. Another possibility is that the two accompanied the rye (*Secale*) which is also well represented. As spade marks were discovered everywhere beneath the plaggen soil cover, the rye may have been dug down. The alternative is that this set of pollen got there by the more common, but still not completely understood, processes which cause pollen to end up in the lower horizons of a natural soil profile.

Unlike in the other two diagrams, buckwheat (*Fagopyrum*), spurrey (*Spergula*), Cruciferae and *Centaurea cyanus* already occur in the B horizon here. This suggests a younger date for the beginning of agriculture on this particular patch of land. Perhaps the rather wet conditions (gley in the subsoil) made this land less suitable for cultivation in the Early Middle Ages.

The plaggen soil itself shows low values for Ericales. Heath seems to have contributed little to the manure. Peat (*Sphagnum* spores) was used, but the main source of the soil accumulation must have been grass turves. This is not immediately apparent from the pollen values, but can be inferred from the granular analysis of the soil. The conclusion of this analysis is that the mineral matter of which the plaggen soil is composed is exactly the same as the original soil and that the turves must have been cut from the same kind of terrain.

The Aanp horizon does not show the increase in Chenopodiaceae pollen visible in the other diagrams. Maize was not discovered either. Clover (*Trifolium*), however, shows an increase. In addition to this, one grain of flax (*Linum usitatissimum*) was counted. The marked increase in Gramineae can be attributed to an increase in the use of grass turves as opposed to heather in the last phase of plaggen manuring, but also to grassy weeds becoming more and more of a problem or to a combination of the two.

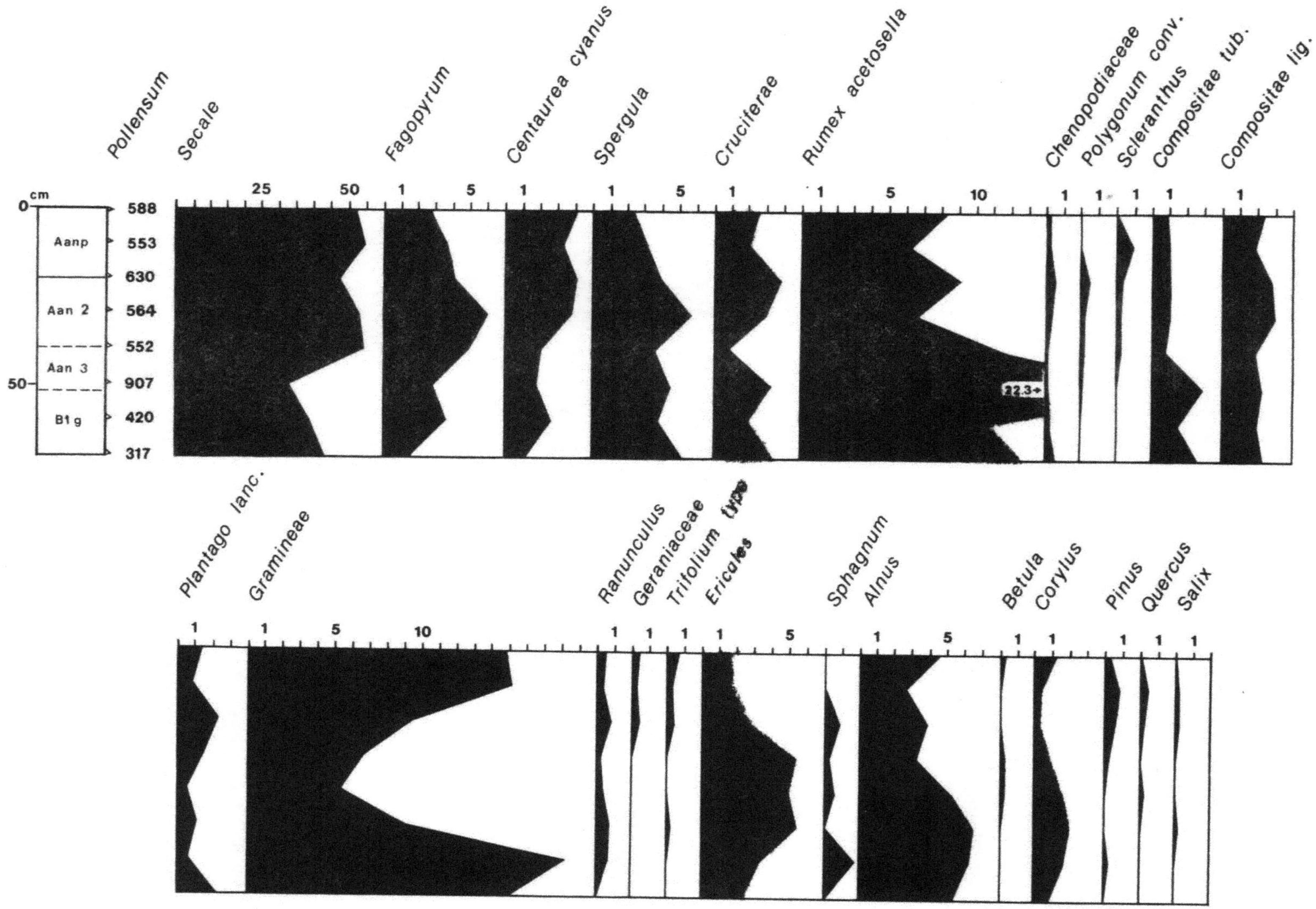

Fig. 4.6 The pollen diagram from Oss-IJsselstraat. Drawing: IPL.

4.3 Conclusions

The problem of working with pollen from plaggen soils is not so much the identification and counting of the pollen grains, but rather the interpretation of the sources of the resulting spectra. The reason for this is that the pollen sources are interwoven to a high degree. The manure used on the fields has the same ultimate source as the local and regional pollen rain. Even the crop plants do not stand out, as the straw (containing pollen) and the plants grown for fodder were eaten by the animals. Nevertheless, a few points can be made.

1. The diagrams confirm that the colour of the plaggen soil is derived from the source of the turves used. According to current opinion, the colour of the black type of plaggen soil is derived from turves of heather. A brown plaggen soil is thought to be the result of grassy turves cut in valleys. The black plaggen soil of Borkel en Schaft indeed shows up to 30% Ericales whilst the brown plaggen soils at Son en Breugel and Oss contain respectively 9% and 5% Ericales at the most.

2. Rye dominates in the pollen spectra. This is in agreement with the historic records, all of which mention rye as by far the most important crop on sandy soils. In the Middle Ages rye was already almost the sole cereal crop (Slicher van Bath 1963). Written sources dating from the 19th century describe the crop rotation as follows: first year rye, second year rye, third year buckwheat, potatoes or something else. Spurrey was sown as an autumn crop after the second year of rye. Other crops from stubble fields were rape-seed, turnips and clover. A record from 1869 for the southern and eastern parts of North Brabant gives a list of the crops grown. The most important crops are given in Table 4.2 together with the corresponding hectares (Table after Trienekens 1985).

Table 4.2 The main crops sown on sandy soils in the year 1869 and their presence in the pollen record

	hectares	in pollen record
Rye	25525	+
Spurrey	13196	+
Potatoes	7190	–
Oats	6091	±
Buckwheat	5102	+
Turnips	3313	?
Carrots	1882	±
Clover	1584	+
Flax and linseed	1153	±
Rape-seed	616	+

+ = present, ± = rare – = absent

If the plants mentioned in Table 4.2 are compared with the pollen record preserved in the Aanp horizon of the plaggen soils, which is thought to embrace the 19th century, it appears that some crops are under-represented or not represented at all. These are potatoes, carrots, oats and flax. The scarcity of Umbelliferae pollen (carrots) can be explained by the fact that a good crop of carrots does not flower. The same can be said of turnips, but their pollen cannot be distinguished from rape-seed. Oat pollen is present in every spectrum but there are few grains. Flax is only present in one spectrum at Oss and potato pollen was not found at all. On the other hand, hemp (Cannabis) is almost always over-represented. In 1869 the plant covered only 9.5 ha.

The reason behind the under- and over-representation might lie in the amount of pollen which the different species produce and disperse. Rye and hemp are good scatterers of pollen. There is, however, another possibility. As mentioned above, the grains of the crop plants may not have fallen down from the crops on the field, but may have got there in the manure. In that case plants producing bedding material or fodder have an advantage over others. It is known that spurrey and clover were grown for the animals. It is also known that these were fed on straw.

3. Cornflower (Centaurea cyanus) is not a Siamese twin of rye, as has been suggested in the past. The facts are in agreement with the findings of Pals and van Geel (1976); cf. also Groenman van Waateringe (this volume, figs. 5-6).

4. Rumex acetosella, sheep sorrel, is a common weed in rye fields. Nevertheless, its pollen curves do not follow the curve of Secale, but behave independently. Another source of sheep sorrel could be patches of heath which were opened when turves were cut there. Anyhow, in both cases the plant is associated with agricultural practices. It is, therefore, not permissible to use the plant as an indication of the presence and the extent of pasture in regions with sandy soils, at least not in the Middle Ages and later periods. This is in agreement with Behre's findings (Behre 1981).

Acknowledgements

I would like to thank Drs. J. Assendorp for the composition of fig. 5, Ir. J. Poelman for the description of the soil profiles and Mrs. S. Mellor for revising the English text.

References

Behre, K.-E., 1981. The interpretation of anthropogenic indicators in pollen diagrams. Pollen et Spores 23, 225-245.

Faegri, K. and Iversen J., 1964. Textbook of pollen analysis, second revised edition. Copenhagen. Munksgaard.

Groenman-van Waateringe, W., 1988. Palynology of plaggen soils on the Veluwe, central Netherlands. In: W. Groenman-van Waateringe and M. Robinson (eds.), Man-made soils. (This volume), 55-65.

Havinga, A.H., 1984. A 20 year experimental investigation into the differential corrosion susceptibility of pollen and spores in various soil types. Pollen et Spores 26, 541-557.

Pals, J.P. and van Geel, B., 1976. Rye cultivation and the presence of cornflower (*Centaurea cyanus* L.). Berichten van de Rijksdienst voor het Oudheidkundig Bodemonderzoek 26, 199-203.

Pape, J.C., 1970. Plaggen soils in the Netherlands. Geoderma 4, 229-255.

Slicher van Bath, B.H., 1963. The agrarian history of Western Europe A.D. 500-1850. London: Arnold.

Trienekens, G., 1985. Het drieslagstelsel en de Vlaamse bouw op de zandgronden van de Meierij: een momentopname voor het begin van de modernisering van de landbouw rond 1860/70. Brabants Heem 37, 2-14.

5. PALYNOLOGY OF PLAGGEN SOILS ON THE VELUWE, CENTRAL NETHERLANDS

W. Groenman-van Waateringe

Abstract

Extensive archaeological research on the Veluwe, central Netherlands, concerning the medieval period, resulted in a programme for pollen analysis of agricultural soils dated from the 7th century until at least the late medieval period. The sites analysed are Kootwijk 2, Hoog Buurlo and Ermelo-Horst.

5.1 Introduction (fig. 5.1)

From 1971-1981 a large excavation and research programme on the Veluwe, centred around the village of Kootwijk was carried out by the IPP (Heidinga 1984a; this volume). The main project involved a large scale excavation of an early medieval deserted village, Kootwijk 2, dated between ca. 750-1000 AD. The village, lying at the edge of a fen pool, started as a cluster of six-eight farmsteads, to become in the 9th century a village consisting of about 20 farmsteads with their barns, sunken huts and wells. In the 10th century the pool dried out (Heidinga 1984a, 1984b) leading to the abandonment of the settlement. The arable was situated to the north and west of the settlement and when the village shifted to the east, the arable took over the sites of the former dwelling-places. Especially in the dry 10th century wind blown sand covered the arable. Heidinga (1984a) assumes that the fields were then raised with humic material from elsewhere mixed with manure to increase the moisture retaining capacity of the topsoil, thus leading to the formation of the oldest plaggen soils yet known in the Netherlands.

The most suitable soils for agricultural purposes in the neighbourhood of Kootwijk were small pockets of loam-rich preglacial deposits with a moderpodzol. The majority of the surrounding soils, however, consists of sands poor in loam and showing a humus podzol (Heidinga 1984a; Pals in press). In addition to the settlement of Kootwijk 2, sites were found from the 2nd and 3rd century, the 6th-7th century, the 7th century and the 11th-12th century.

A small trial excavation was carried out at Hoog Buurlo, ca. 5km to the SE of Kootwijk. Here, traces of a small settlement dated to the 6th-7th century were found. Conditions for agriculture must have been very marginal, since the soils were so poor and dry. The comparatively wealthy condition of the inhabitants (the import of Frankish pottery being particularly high) is explained by Heidinga (1984a) in terms of industrial activity, with iron ore processing and exporting on a wide scale.

Fig. 5.1 Situation of the sites mentioned in the text: 1, Kootwijk; 2, Hoog Buurlo; 3, Ermelo-Horst. Drawing: IPP.

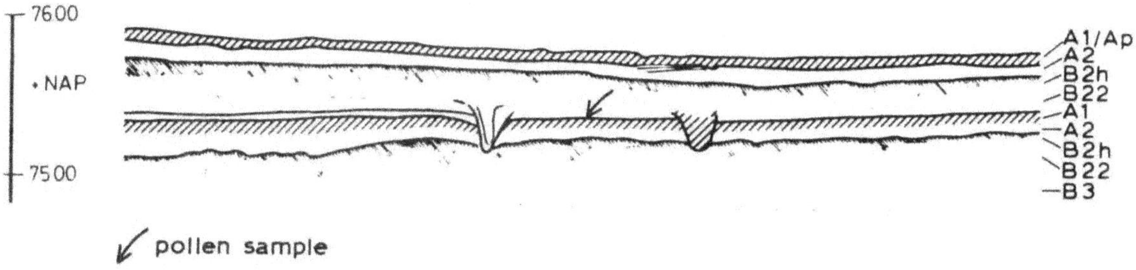

Fig. 5.2 Kootwijk 2, section through agricultural layers NE of the former pool with location of pollen samples. After Heidinga 1984a, fig. II, 1:11. Drawing: IPP.

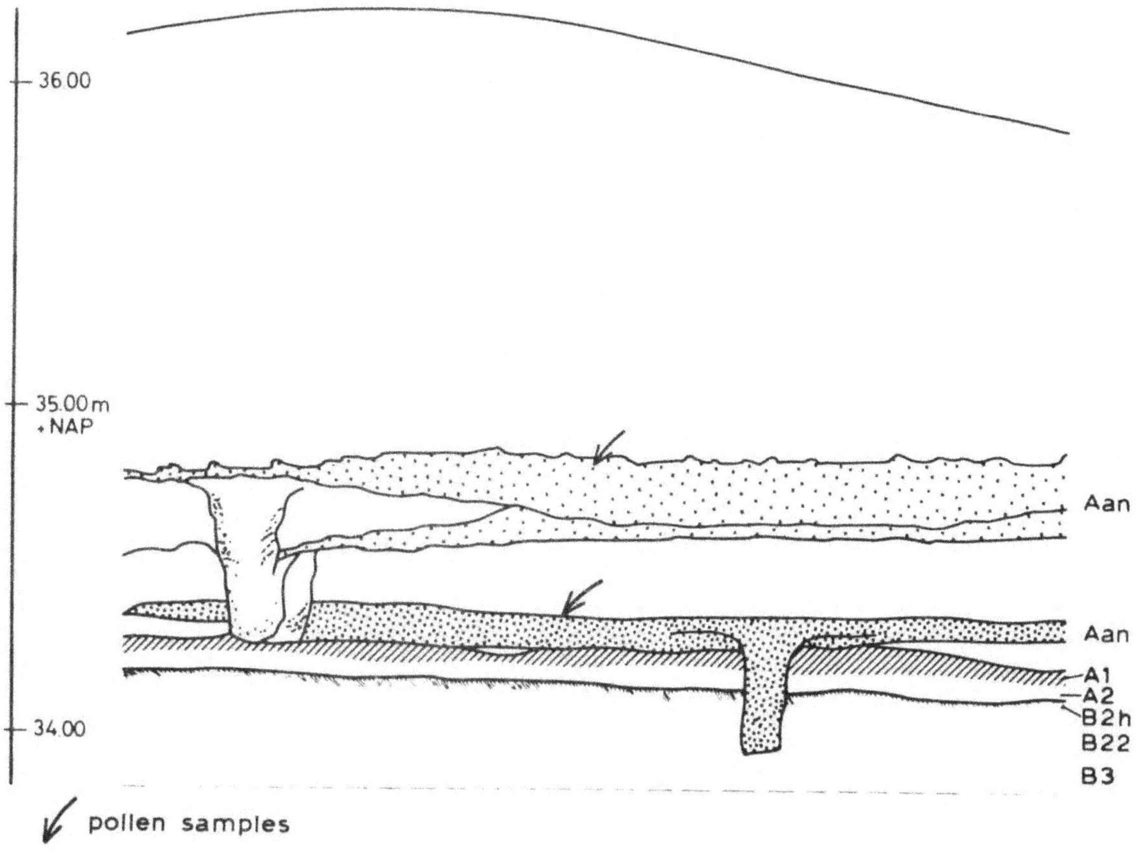

Fig. 5.3 Hoog Buurlo, section of the blown-sand S of the settlement with location of pollen sample. After Heidinga 1984a, fig. II, 4:5. Drawing: IPP.

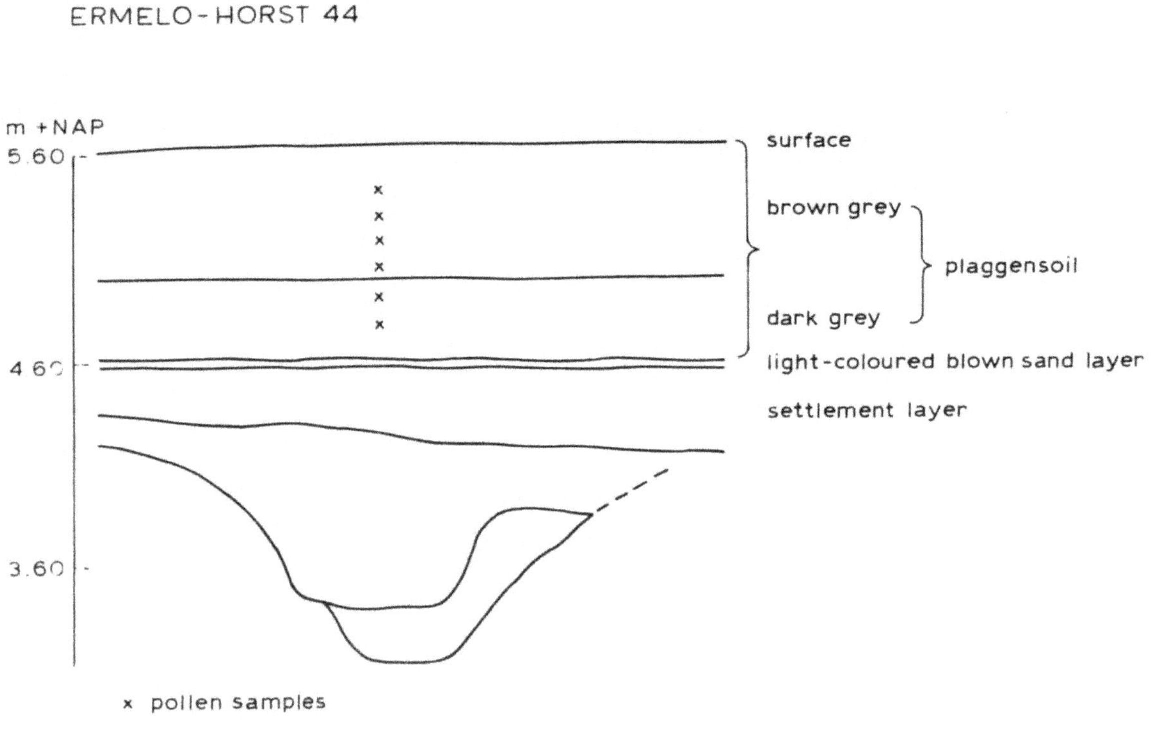

Fig. 5.4 Ermelo-Horst, section through ditch, 12th century settlement layer and plaggen soil with location of pollen samples. Drawing: IPP.

A rescue excavation was carried out in the hamlet of Ermelo-Horst, where a settlement of the 12th century was covered by a late medieval plaggen soil. The site is situated in a lowland area in the western edge of a coversand.

5.2 Sampling (figs. 5.2-4)

In the Kootwijk site no. 2, samples for pollen analysis were taken from the filling of the wells, the agricultural layers covered by blown sand, from turves in the walls of sunken huts and from the fillings of sunken huts (see van Geel and Groenman-van Waateringe in press). Pollen preservation was mostly rather bad, save in the wells. Here we will only deal with a sample taken from an agricultural layer which Heidinga (1984a, 91ff.) supposed to be an artificially raised arable soil. One sample from a well is included, the preservation condition for pollen in wells being comparable with that in plaggen soils. The samples were analysed by Dr. B. van Geel, Vakgroep Bijzondere Plantkunde, University of Amsterdam. (Full analyses are to be found in van Geel and Groenman-van Waateringe op. cit.).

In Hoog Buurlo samples were taken from a section S of the settlement, where a settlement, or possibly arable, layer, lies on a humus podzol formed in a blown sand, the latter probably to be dated to the Merovingian period. Here, only the sample from the 7th century settlement layer is given.

In Ermelo-Horst samples were taken from the 12th century settlement layer and the plaggen soil on top of it.

5.3 Results (figs. 5.5-6)

In figs. 5.5-6 only the main curves characteristic of cultivation on plaggen soils are given, although full data are available. To allow for a better comparison with the pollen data of plaggen soils published by Behre (1976, 1980), the sum of arboreal pollen formed the basis for percentage calculations.

The state of preservation of the pollen in the sample from Hoog-Buurlo was rather poor. The AP/NAP ratio was 27/73, indicating an open landscape characterised by a grassy heather (Ericaceae/Gramineae ratio 53/47). This would have made it possible to graze sheep and cattle here. The cutting of heather sods, which would have led to impoverishment and thus to a practically pure stand of heather, does not seem to have been practised on a large scale. Six pollen grains of cereals, one of which is a Secale, point to cereal cultivation locally, notwithstanding the difficult circumstances.

Conditions of preservation in the 10th century agricultural layer at Kootwijk 2 were poor, the amounts of pollen and the number of herb 'species' are low in comparison to the late medieval plaggen soil and samples from the wells, and in the same order of magnitude as in the sample from Hoog Buurlo. The total amount of cereal pollen is also low. Secale was definitely identified and according to Pals and van

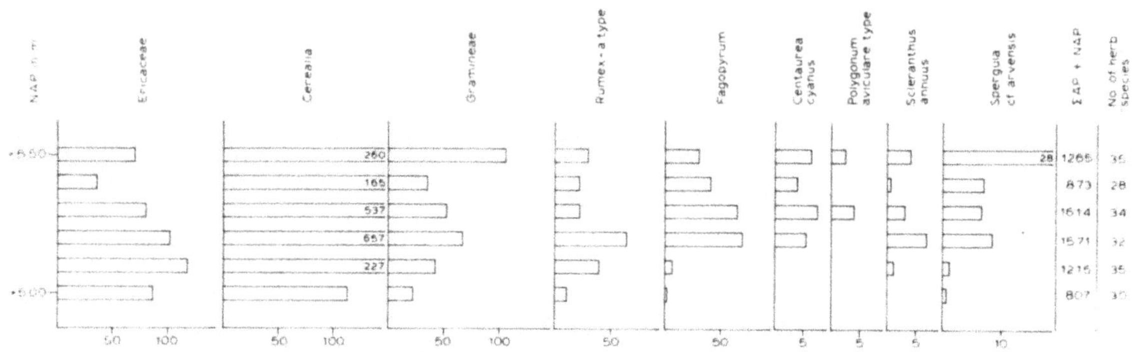

Fig. 5.5 Some pollen data of Hoog Buurlo, Kootwijk 2 well and agricultural layer, and Ermelo-Horst 12th century settlement layer and plaggen soil. Drawing: IPP.

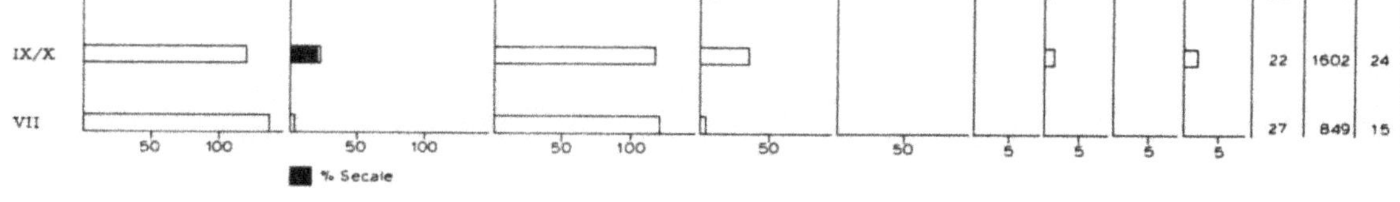

Fig. 5.6 Selection of pollen data of the plaggen soil of Ermelo-Horst. Drawing: IPP.

Geel (1976) there is enough evidence for rye cultivation, although several of the species characteristic for the Secalietea are absent. The AP/NAP ratio was 31/69, comparable to that from Hoog Buurlo. The Ericaceae/Gramineae ratio was 36/64, as is the case in other agricultural layers (cf. Pals and van Geel op. cit.). The presence of a Sordariaceous fungal spore may point to the use of animal manure (van Geel and Groenman-van Waateringe in press; van Geel et al. 1981). The sample from the well was rather better preserved as is reflected in the total amount of pollen counted and the number of herb 'species'. The AP/NAP ratio was 22/78, the Ericaceae/Gramineae ratio 52/48. The 12th century layer of Ermelo-Horst is apparently an agricultural layer, with very high values for the cereals, among which a high percentage for Secale (82%). The AP/NAP ratio is 8/92, the Ericaceae/Gramineae ratio 76/24. This is quite inconsistent with the interpretation as an agricultural layer and is probably caused by the ploughing-up of soil rich in Ericaceae pollen in the reclamation phase. In another sample from the same settlement layer, taken from a section situated more to the west of the settlement with only a small amount of cereal pollen, and apparently not an agricultural layer, the Ericaceae/Gramineae ratio is about the same, 82/18, indicating practically pure heather around the settlement and thus also the intensive use of it.

The samples from the late medieval plaggen soil of Ermelo-Horst reveal firstly that, because of the highly humic nature of the soil and its low pH (cf. Pape 1970, Table 1), the preservation of the pollen is excellent, and, secondly, that for the same reasons the samples are very rich in pollen. Here species characteristic for the Secalietea as Rumex acetosella, Scleranthus annuus, Spergula arvensis, Polygonum aviculare and Centaurea cyanus were present (see figs. 5.5-6), but also Polygonum persicaria, Trifolium type, Viola sp. occurred regularly.

Crop plants consisted in the first place of rye, followed by buckwheat. Triticum type pollen was found regularly, only rarely Hordeum and Avena type and furthermore Linum usitatissimum. Pollen of the family Cruciferae, Vicia and Trifolium type pollen occurred regularly, although in very small numbers and are perhaps evidence for other crops or green manure.

5.4 Discussion and conclusions (Plates 5.I-II)

Comparison of the results of the presumed plaggen soil at Kootwijk with those not only from the late medieval plaggen soil of Ermelo-Horst, but also with the 10th century plaggen soil of Dunum, West-Germany (Behre 1976, 1980), reveals great differences, not so much in the taxa, as in the quantity of pollen and its state of preservation. The humic nature and low pH of a real plaggen soil provide excellent conditions for the preservation of the pollen in it. This certainly was not the case in the arable layer of Kootwijk. Heidinga's argument for an artificially raised arable soil is based upon its situation on top of a blown sand (see fig. 3.3 and Plate 3.I), both layers indicated with Aan. My interpretation of these layers is that they are a normal Ah layer, ploughed and manured (cf. the spore of a Sordariaceaous fungus). The patches of blown-sand which are still visible here and there underneath the lowermost layer (Aan) can be explained as not having been touched by the plough. The same holds true for the

Plate 5.I Hoog Buurlo, section S of the settlement, showing the stratigraphy of two superimposed humus podzols divided by a blown-sand layer. Photo: IPP.

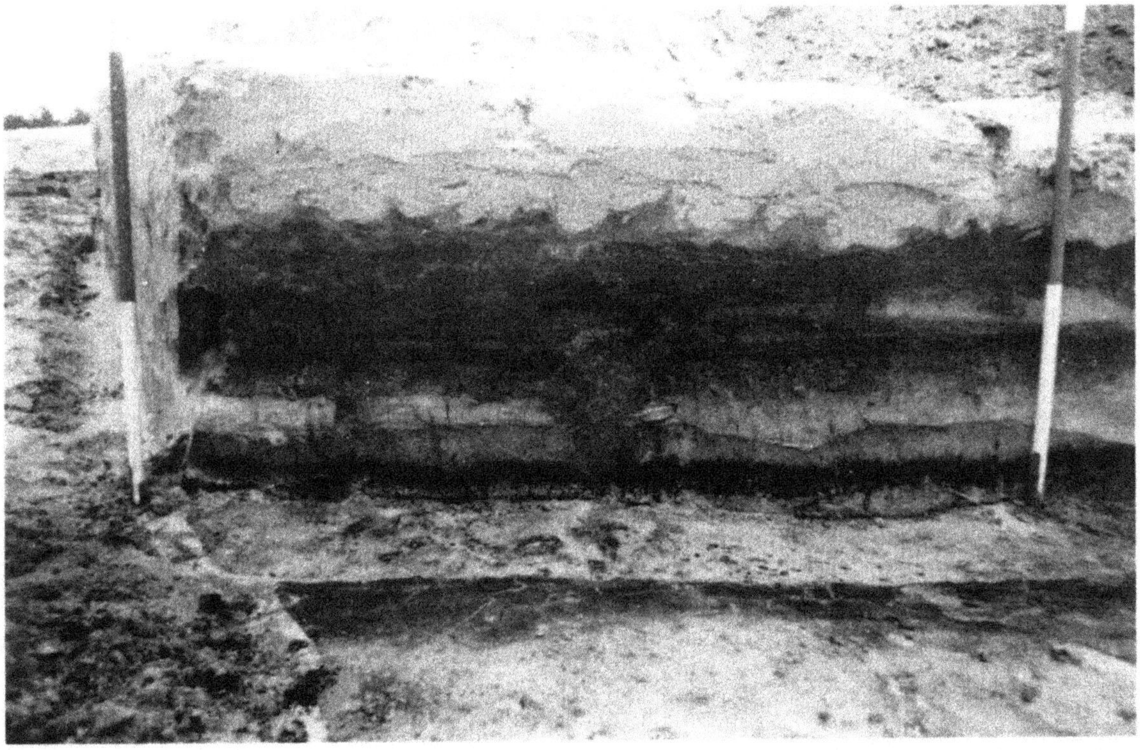

Plate 5.II Kootwijk 2, section through agricultural layers NE of the former pool. Photo: IPP.

uppermost (Aan) layer, which represents renewed cultivation after the earlier fields had been covered by sand. The thickness of the layer (more than 40cm) can be explained by repeated sand drifts covering the fields in such a way that the Ah layer grew in thickness together with the deposition of blown sand. This also explains the light colour of the layer as it is a mixture of humic material and blown sand. Manuring must have been absolutely essential under these circumstances, not only bringing in the required nutrients, but also retaining a high amount of moisture, thus preventing the topsoil with the seed being blown away with the first strong wind (cf. Groenman-van Waateringe 1979). There are, however, no indications in Kootwijk 2 for a deliberate raising of the fields with plaggen or other humic material, such as in Dunum or Ermelo-Horst. That difference in preservation conditions is a major factor here can be seen by comparing the spectrum from the well sample with that from the agricultural layers. In the well the amounts of pollen and the number of herb 'species' are as high as in the plaggen soils. If manuring of the plaggen had taken place in Kootwijk in the way it was done later (Pape 1970; van de Westeringh this volume), the plaggen soaked in dung and urine, would have provided better conditions for the preservation of the pollen.

The pollen content of plaggen soils reflects not only the crops grown on the fields, but also the plants in the plaggen brought with dung onto the fields as well as the composition of the fodder of the dung producing animals (cf. Groenman-van Waateringe in press). Plaggen can come from anywhere, heaths, meadows, brook valleys etc. (cf. Bakels this volume), but also other kinds of material were used, e.g. forest litter, straw etc. Thus the pollen content of plaggen soils can be a mixture of pollen from quite different habitats. The pollen of cereals other than <u>Secale</u> could have been brought in with the straw and it is possible that this was the case with the <u>Triticum</u>, <u>Avena</u> and <u>Hordeum</u> type pollen, because it was known at least from the 19th century that these crops do not thrive well on plaggen soils (Tiesing 1974). The same holds true for <u>Linum usitatissimum</u>, which needs a much more loamy soil, and <u>Fagopyrum esculentum</u>, that was mainly cultivated on peaty soil, although the cultivation of buckwheat on plaggen soils is known. Buckwheat straw was, moreover, very necessary for the rye crop. The lack of buckwheat straw in the 19th century was said to be the reason for diminishing yields. Pollen of plants from wet habitats found in samples from Ermelo-Horst, like <u>Lythrum</u>, <u>Thalictrum</u> and <u>Filipendula (ulmaria)</u> could, via hay eaten by the animals during winter stabling, finally turn up in the plaggen soils.

Acknowledgements

Thanks are due to Dr. Carol van Driel-Murray for improving my English, Dr. Anthonie Heidinga for discussions and permission to publish figs. 5.2-3 and Plates 5.I-II, Dr. Bas van Geel for allowing me to use two of his analyses from Kootwijk 2 here, Cees Troostheide for preparing the pollen samples, Fred Gijbels for the photos, Bob Donker for the drawings 5.1 and 5.4-6 and Trudy Domhoff for typing the text.

References

Bakels, C.C., 1988. Pollen from plaggen soils in the province of Northern Brabant, the Netherlands. In: W. Groenman-van Waateringe and M. Robinson (eds.), (this volume), 35-54.

Behre, K.-E., 1976. Beginn und Form der Plaggenwirtschaft in Nordwestdeutschland nach pollenanalytischen Untersuchungen in Ostfriesland. Neue Ausgrabungen und Forschungen in Niedersachsen 10, 197-224.

Behre, K.-E., 1980. Zur mittelalterlichen Plaggenwirtschaft in Nordwestdeutschland und angrenzenden Gebieten nach botanischen Untersuchungen. In: H. Beck, D. Denecke, H. Jankuhn (eds.), Untersuchungen zur eisenzeitlichen und frühmittelalterlichen Flur in Mitteleuropa und ihrer Nutzung, Tl. II. Abhandlungen der Akademie der Wissenschaften in Göttingen, Phil.-Hist. Klasse, 3. Folge, 116, 30-44.

Faegri, K. and Iversen, J., 1964. Textbook of pollen analysis. Copenhagen: Munksgaard.

Geel, B. van, Bohncke, S.J.P. and Dee, H., 1981. A palaeoecological study of an upper Late Glacial and Holocene sequence from 'De Borchert', the Netherlands. Review for Palaeobotany and Palynology 31, 367-448.

Geel, B. van and Groenman-van Waateringe, W., 1987. Palynological investigations. In: W. Groenman-van Waateringe and L.H. van Wijingaarden-Bakker (eds.), Farm life in a Carolingian village. Studies in Prae-en Protohistorie 1. Assen: van Gorcum and Comp, B.V.

Groenman-van Waateringe, W., 1979. Nogle aspekter af Jernalderens agerbrug i Holland og Nordvesttyskland. In: H. Thrane (ed.), Fra Jernalder til Middelalder. Skrifter fra Historisk Institut, Odense Universitet 27, 75-86.

Groenman-van Waateringe, W., 1986. Grazing possibilities in the Neolithic of the Netherlands based on palynological data. In: K.-E. Behre (ed.), Anthropogenic indicators in pollen diagrams. Rotterdam: A.A. Balkema, 187-202.

Heidinga, H.A., 1984a. De Veluwe in de vroege middeleeuwen. Aspecten van de nederzettingsarcheologie van Kootwijk en zijn buren. Thesis Amsterdam.

Heidinga, H.A., 1984b. Indications of severe drought during the 10th century AD from an inland dune area in the central Netherlands. Geologie en Mijnbouw 63, 241-248.

Heidinga, H.A., 1988. Climate and plaggen soils. In: W. Groenman-van Waateringe and M. Robinson (eds.), Man-made soils. (This volume), 21-33.

Pals, J.P., 1987. Reconstruction of landscape and plant husbandry. In: W. Groenman-van Waateringe and L.H. van Wijngaarden-Bakker (eds.), Farm life in a Carolingian village. Studies in Prae-en Protohistorie 1. Assen: van Gorcum and Comp, B.V.

Pals, J.P. and Geel, B. van, 1976. Rye cultivation and the presence of cornflower (Centaurea cyanus L.). Berichten van de Rijksdienst voor het Oudheidkundig Bodemonderzoek 26, 199-204.

Pape, J.C., 1970. Plaggen soils in the Netherlands. Geoderma 4, 229-255.

Tiesing, H., 1974. Over landbouw en volksleven in Drenthe. Dl. I. Verzameld en samengevat door Dr. C.H. Edelman. Assen: van Gorcum and Comp. B.V.

Westeringh, W. van de, 1988. Man-made soils in the Netherlands, especially in sandy areas (plaggen soils). In: W. Groenman-van Waateringe and M. Robinson (eds.), Man-made soils. (This volume), 5-19.

6. ARCHAEOLOGICAL AND ENVIRONMENTAL ASPECTS OF COLLUVIATION IN SOUTH-EAST ENGLAND

M.J. Allen

Abstract

Over a decade of study of colluvial deposits in lynchets and valley locations has provided a sound foundation for a new research impetus to be directed at such material.

Here I wish to introduce new methods of study, including archaeomagnetic dating, magnetic susceptibility and detailed micromorphometry and indicate how man has inadvertently modified his landscape. Further, from these man-made, or man induced, colluvia I wish to show the degree of environmental interpretation that can be gained from their study detailing work from two sites in Sussex; Ashcombe Bottom, Lewes and Bourne Valley, Eastbourne. The detailed investigations from these sites and others studied by Bell and Kerney now allow a regional interpretation to be sought and integrated with the strictly archaeological data.

6.1 Introduction

Recent studies of colluvial dry valley sediments have provided both environmental and anthropogenic evidence for their occurrence (e.g. Kerney et al. 1964; Evans 1966; Evans and Valentine 1974; Bell 1982; 1983). Detailed archaeological investigations in SE England (fig. 6.1) by Bell (1981a; 1982; 1983) and Allen (1983; 1984a) have shown that colluvia in valley bottoms contain large quantities of stratified archaeological artefacts (Table 6.1) which indicate the high level of anthropogenic activity and exploitation of the landscape and contradict climatic interpretations placed on Mediterranean fills (Bintliff 1977) and alluvial deposits (Potter 1976). Studies by Bell (1981a and b; 1982; 1983), Allen (1983; 1984a) and Ellis (1986) have shown a widespread occurrence of colluvium in valley floors on the chalklands of SE England. This work provides a basis for a new research impetus to be directed at colluvium because it can now be assumed:

a) that there is large scale occurrence of colluvium,

b) that such material contains archaeological artefacts,

and c) that colluvia are broadly anthropogenic in origin.

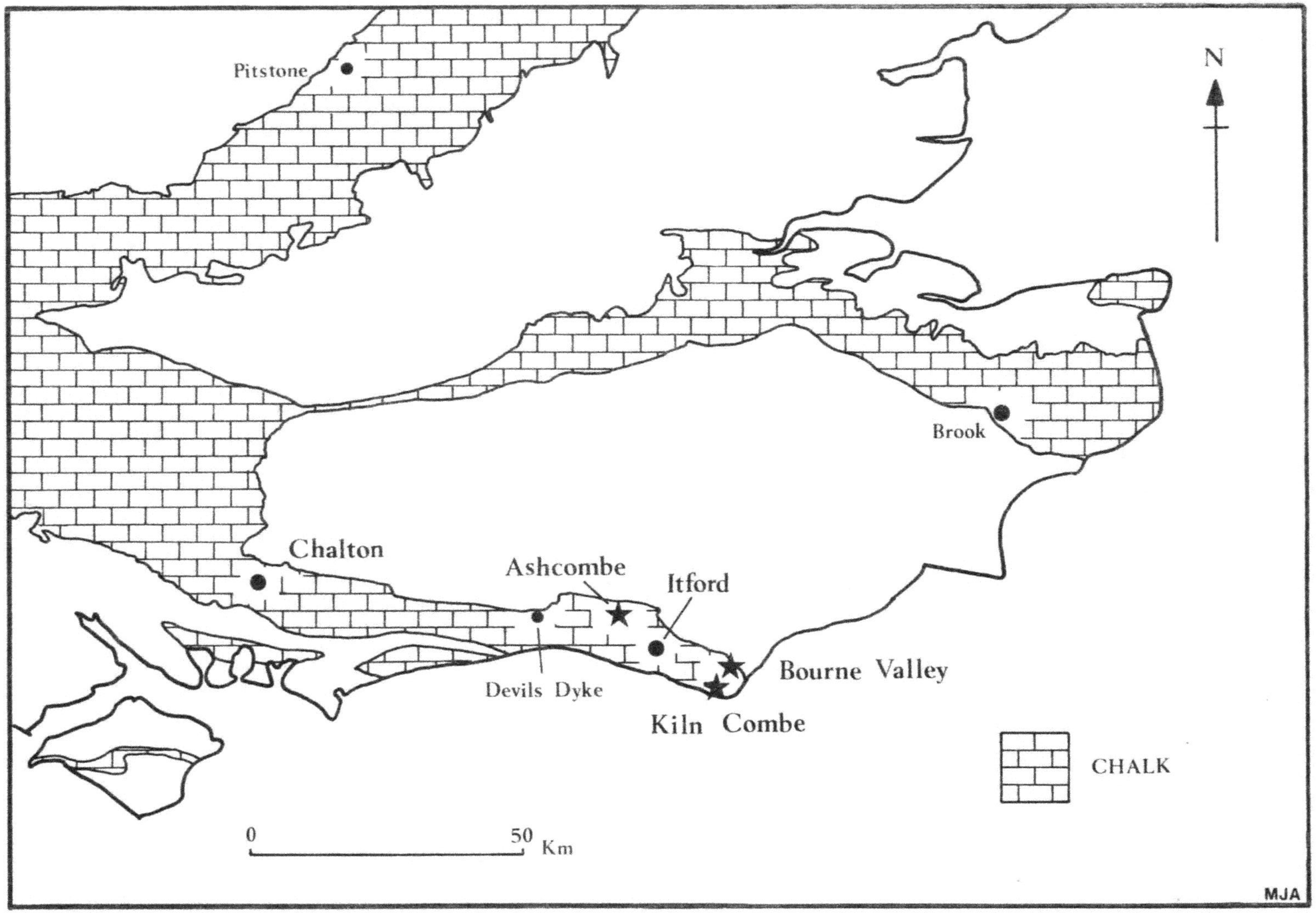

Fig. 6.1 Location of study areas mentioned in the text; stars represent archaeological sites beneath colluvium.

	a	b	$\frac{a}{b}$ =			
	Estimated m^2 of colluvium in whole cross section	Length of contributing slope in metres	Estimated depth of eroded soil in cm.	Total artifacts excavated	Artifacts per m^3 of hand-excavated soil	Date of Erosion
ASHCOMBE BOTTOM Valley TQ 380 106	24	500 m	5 cm	1911	138·1	Beaker → modern
BOURNE VALLEY Lynchet TV 600 994	156	640 m	24 cm	c. 10,000	256·4	Bronze Age → Saxon
KILN COMBE Valley TV 573 964	90	500 m	18 cm	3109	68·2	Beaker → Medieval
BISHOPSTONE Lynchet TQ 468 006	15·4	160 m	10 cm	1985	64	Neolithic → Saxon
ITFORD BOTTOM Valley TQ 441 049	59	1,200 m	12 cm	2103	68·9	Early Bronze Age → Roman
CHALTON Valley B SU 729 160	60	700 m	9 cm	3105	204·3	B.A./I.A. → Modern
CHALTON Valley A SU 721 166	22·6	900 m	2·5 cm	366	114·3	Neolithic → Modern
HAMBLEDON Spur	c. 30·6	500 m	6 cm	163	—	? Prehistoric → Modern
HAMBLEDON Coombe	c. 137·6	680 m	20 cm	167	104·4	? Iron Age → Modern

MJA

Table 6.1 An attempt to quantify (a) the extent of erosion represented by surviving colluvium, and (b) the numbers of archaeological artefacts in the excavated sections. The final column provides a crude indication of the period over which erosion occurred. (After Bell, 1986).

This basic framework will allow us to refine and explore further the potential of these man-made or man induced deposits both site specifically and regionally. It is becoming increasingly evident that the soils on the chalklands, like many other areas, owe much of their present characteristics to anthropogenic activity. The grey rendzinas seen on the Downland today and the thick colluvial blankets in valley bottoms are a product of man, moreover both of these units contain quantities of pottery and charcoal proving their strong anthropogenic affinities.

Bell (1982, 88) states "much remains to be done to realise the full potential of valley sediments in palaeoenvironmental studies... and... Many problems of chronology and causation require much refinement". Bell does not, however, consider the larger scale, regional pattern of environmental exploitation which can be evaluated from the corpus of work directed at hillwash. Here I wish to address both the specific problems of causation and of regional environmental reconstruction. The author's research has been specifically directed at:

a) the dating of the sediments,

b) the environments in which and from which the sediments erode,

c) the nature and causation of erosion,

d) the pedological implications,

and e) the archaeological and sedimentological implications of these deposits.

6.2 Dating the deposits

Bell (1981b; 1982; 1983) obtained a chronology of colluvial sequences by examining the distribution of datable artefacts within the sediments and also radiocarbon determinations of associated charcoals were obtained. From such distributions it has been demonstrated that colluvial sequences are chronologically stratified (e.g. fig. 6.2) and thus provide a basis for palaeoenvironmental data as well as hinting at the mode of sedimentation and lack of gross disturbance on deposition. However, the materials Bell used to provide this chronology, pottery and radiocarbon determinations, are extrinsic components of the sediments and may themselves suffer residuality - that is to say that there may be a gross temporal hiatus between artefact 'loss' and subsequent incorporation into the colluvial sedimentary unit, especially as archaeological field survey readily provides evidence of prehistoric pottery scatters on hill slopes and valley sides. This indicates that the material has not been totally re-worked into valley bottoms and that artefacts of great antiquity are still eroding from valley sides at present. The fact that chronologically sensible artefact distributions (fig. 6.2) have been present in all the valleys investigated so far argues against gross distortion due to residuality. None the less, the dating of extrinsic artefacts is bound to provide a relatively crude chronology and it is important to explore methods of dating the sediments on the basis of their own intrinsic properties.

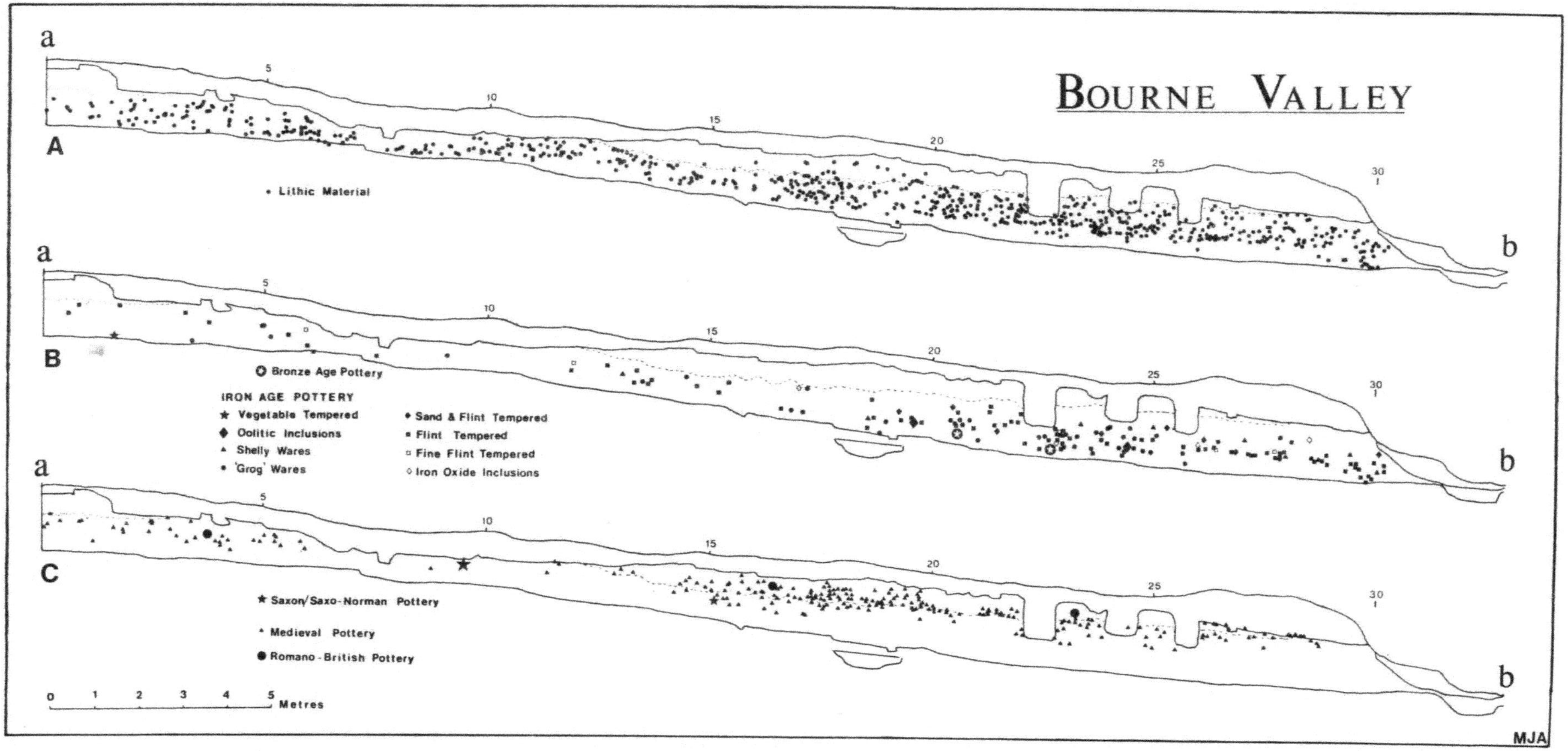

Fig. 6.2 Distribution of artefacts in the Bourne Valley colluvial section.

Experimentation with archaeomagnetic dating at the Bourne Valley, Sussex has provided a tentative glimpse of how an intrinsic component, orientated magnetic domains, may provide an alternative dating method (Allen 1983). Archaeomagnetic dating (or palaeomagnetic secular variation) records that the depositional remnant magnetism (DRM) of domains within the sediment that have been preferentially orientated to magnetic north at the time of, or shortly after, deposition. This orientation is fossilized and can be measured (Tarling 1971; 1975; Clark 1980) using a balanced fluxgate magnetometer.

The colluvial deposits at the Bourne Valley, Sussex (TV 600 994) contained evidence for extensive Iron Age activity, but little ceramic evidence for the Romano-British period was encountered (fig. 6.2). Archaeomagnetic determinations made at the Geophysics section of the Ancient Monuments Laboratory (H.B.M.C.), however, provided evidence for continuing sediment deposition from the Iron Age and throughout the Romano-British period. Colluviation at this site would have been seriously arrested in the medieval period by the construction of a road or drove-way upslope. The archaeomagnetic dates therefore correlate well with the artefact distribution. Moreover, the lack of Romano-British material may be seen as an 'artefact' of manuring techniques or land usage, rather than a real hiatus.

Using a combination of both artefact distribution and archaeomagnetic dating a more refined chronology may be obtained. The basic fact that the colluvia display magnetic orientation also allows statements to be made about the mode of sediment deposition. The colluvium must have been eroded, transported and deposited in such a way that the sediments were 'lubricated' by agencies of wind or water allowing the magnetic domains to become preferentially orientated to the magnetic north at the time of deposition. This has direct implications upon erosional and depositional regimes, as discussed later.

6.3 Methods of analysis

6.3.1 Magnetic Susceptibility

Magnetic susceptibility is familiar to archaeologists as a method of geophysical prospecting, detecting burning episodes attributed to man (Tite and Mullins 1970; 1971; Mullins 1974; Aitken 1974; Longworth and Tite 1977; Clark 1983). However, geographers have demonstrated its use in lacustrine sediments as a palaeoenvironmental indicator allowing distinctions to be made between sediments derived from topsoil or subsoil and also from different ecological/vegetational regimes, i.e. woodland, grassland and arable (Dearing et al. 1981; Oldfield et al. 1978). The potential that such interpretive statements may also be made of terrestrial sequences has been briefly explored by Allen (1983) and Allen and Macphail (1985; 1987).

Magnetic susceptibility (χ_0) is the ratio between the magnetisation induced in a sample and the magnetising field (detailed definitions are given in Thompson et al. 1975 and Oldfield et al. 1978). Magnetic susceptibility <u>enhancement</u> is due to the conversion of iron oxides in the soil from weakly ferro-magnetic forms, e.g. haematite

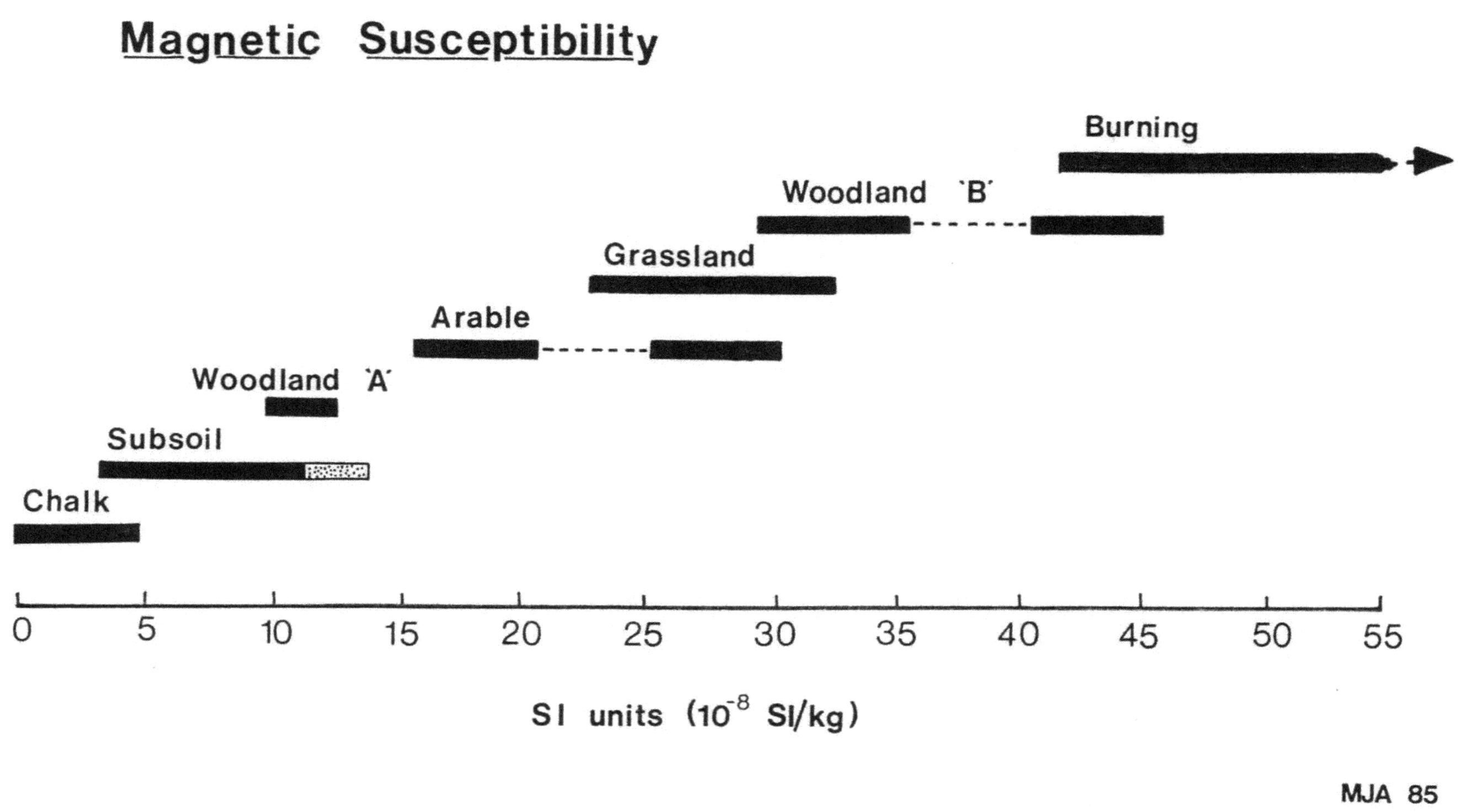

Fig. 6.3 Summary diagram of magnetic susceptibility variation under modern chalkland regimes. (Compiled from ca. 1,000 readings from the Sussex Downs).

(αFe_2O_3), to strongly magnetic forms, e.g. geothite and maghaemite. This enhancement may be due to burning (Clark 1983) or pedogenic processes (Allen 1983; Allen and Macphail 1985; 1987).

Over 1,000 magnetic susceptibility measurements from modern environments seem to indicate that on the chalklands, which have 'low background magnetic susceptibility', soils may be differentially enhanced according to their vegetational and pedogenic regime. Woodlands in general seem to produce higher magnetic susceptibility values than secondary woodland or scrub, which in turn produce higher values than grassland, arable and subsoils respectively (fig. 6.3).

A suite of 105 magnetic susceptibility determinations using a Bartington MS 1 meter coupled to a MS 1B sensor coil calibrated for 100g of soil, taken from the Bourne Valley show that the sequence consists of topsoil (fig. 6.4), and further more may equate to magnetic susceptibility levels in arable environments (Allen 1983). This hypothesis complements other environmental data from the site, e.g. sedimentological, micromorphological and molluscan, which suggest that the sequence may be a lynchet, albeit truncated (Allen 1983) in an 'alluvial edge' location (cf. Bell 1981b, 76).

If magnetic susceptibility values are preserved in archaeological situations then we have, potentially, an additional palaeoenvironmental technique to add to the growing suite of determinants. Magnetic susceptibility enhancement seems to be preserved in the Bronze Age buried grassland soil at North Kite, Wiltshire. Here magnetic susceptibility values for the buried soil, which showed typical grassland characteristics (heavily worm sorted and contained the open grassland Mollusca (Evans 1972) Trichia hispida, Helicella itala and Vallonias), were almost identical to present day magnetic susceptibility values from grassland in the same location. However Allen and Macphail (1985; 1987) have shown that the post burial taphonomic processes, such as localised hydromorphism and redox, may both deplete and concentrate iron in soil horizons. This produces a variable pattern of magnetic susceptibility values differing from the original archaeological levels. Furthermore, complicated pedogenic histories may result in the inclusion of lithorelicts, clay illuviation and hydromorphism which will produce anomalous magnetic susceptibility readings. Some of these problems can be resolved using microfabric analysis and indeed the combination of these two techniques has been shown to be useful in both palaeoenvironmental reconstruction and the understanding of complex pedogenic histories (Allen and Macphail 1985; 1987). Here I have only briefly indicated the potential and pitfalls of magnetic susceptibility as a palaeoenvironmental indicator, however further research into this field is obviously needed.

6.3.2 Micromorphology

Micromorphological analysis (the microscopic examination of physical soil characteristics in thin section) of colluvial deposits in Western Europe by Mücher (1974) and Kwaad and Mücher (1979) has enabled the differentiation between soil and sediment to be made. Moreover, they have been able to isolate cultivation episodes as well as elucidate a complex landscape history by analysis of clay coatings and

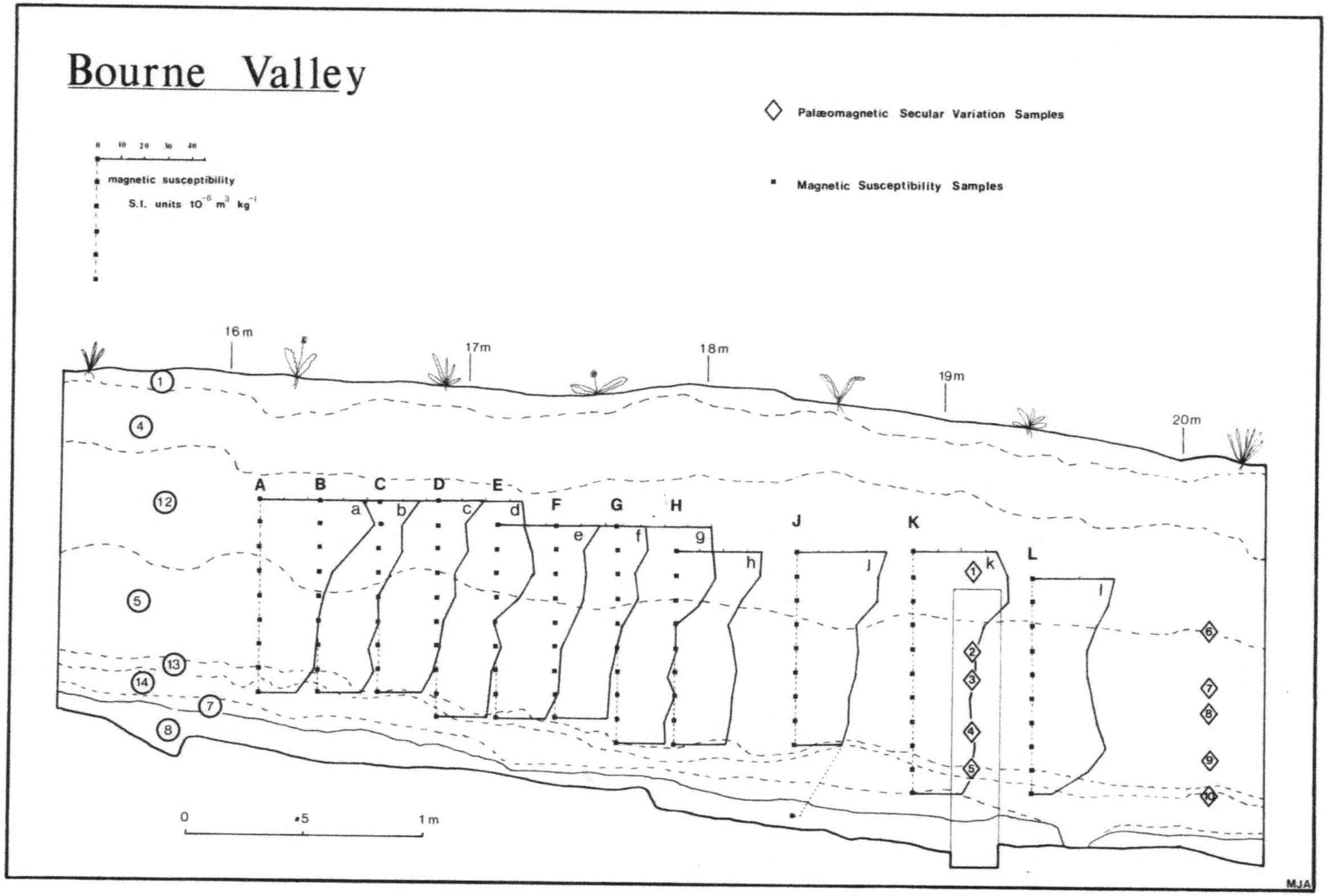

Fig. 6.4 Magnetic susceptibility; sample location and values from the Bourne Valley.

mineralogenic coatings. Although Kwaad and Mücher showed the enormous potential for palaeoenvironmental investigations via microfabric analysis, little work has been undertaken on colluvial deposits in England. Macphail has provided some information from limited studies on the colluvial sequences at Chalton (Macphail 1983) and also the Bourne Valley where he showed secondary chalky infills within the colluvium related to the Iron Age and Roman cultivation (Macphail et al. 1987) thus demonstrating the usefulness of employing this technique on fine grained calcareous colluvial sediments. Fisher and Macphail (1985, 93) have recently illustrated the importance of microfabric analysis in providing "accurate information about past environments and anthropogenic activity". Furthermore Hall's (1987) micromorphological work on modern arable and colluvial contexts has allowed him to identify specific erosion fabrics which may be directly identifiable in past erosional regimes. From this brief evaluation it is clear that the potential for micromorphological studies has not been fully realised and if we are really to attempt to understand the processes of colluviation and evaluate the palaeo-landscape and past land-used history then this technique must be applied more rigorously to such deposits.

Using the techniques of magnetic susceptibility, micromorphology and archaeomagnetic dating some of the problems outlined by Bell (1982, 88) - the chronology and causation of erosion - can be tackled, allowing a higher degree of interpretation. However, one major and crucial problem is that of the exact nature and mode of erosion that produces quantities of sediment in valley and lynchet locations. This is of paramount importance to the understanding of the deposits and the potential taphonomic effects of their archaeological and biotic artefacts.

6.4 Erosion regimes

It is becomining increasingly obvious that studies of modern events are needed as analogues for past studies and that archaeologists and geographers have often underestimated the size of the erosion events they are dealing with. Too much emphasis may have been placed upon the processes of rainsplash, seasonal soil creep and ploughwash and less on larger scale more infrequent erosion events. Erosion is an ongoing process which, as Kwaad and Mücher (1977) have shown, occurs naturally to a degree even under heavy woodland. Natural erosion is minimal in comparison with man-induced erosion and rates of erosion are greatly accelerated by devegatation i.e. for tillage and other anthropogenic activities (Dimbleby 1976). Bell (1981b) and Kwaad and Mücher (1979) have tended to place emphasis on the movement of material downslope by the plough, which has allowed Bell to argue for the similarity in origin of deposits located in lynchets and valley bottoms. Although Bell has redressed his ideas (Bell 1983; 1986) the importance of larger scale erosion events and the subtle difference between valley deposits and lynchets still need stressing. In both cases it is evident that causation ultimately remains the same, i.e. anthropogenic activity, but the scale of event may vary.

6.4.1 Lynchets

The lynchets studied at Bishopstone, Sussex TQ 468 006 (Bell 1977), Bullock Down, Sussex TV 587 973 (Clarke 1983) and the Bourne Valley (Allen 1983) are all characterised by highly calcareous silty clays/silty clay loams with many small, well rounded chalk pieces and only a few larger components. The deposits are fairly uniform, perhaps indicating that the sequences had accumulated steadily. The nature of the deposits indicates low energy erosion, probably originating within a single field unit; thus we can see an 'intra-field' erosion regime operating whereby slope length is restricted by field size. 'Celtic' field size itself is possibly governed by time-and-motion logistics, i.e. the area that can be ploughed by one team in one day (Bowen 1961). Groenman-van Waateringe (1979) has suggested that some 'Celtic' fields in North Western Germany and Holland represent attempts to counter intra and inter-field erosion. The quantity of chalk pieces in the Sussex lynchets indicates a lack of soil depth at some locations upslope, although not necessarily ubiquitously over the whole field. It should be noted though, that the archaeological sections through the lynchets mentioned are only a very small proportion of the whole feature and thus any evidence of localised large scale higher energy erosion events probably remain unsampled. However it is likely that occasionally 'lynchet-spill' will occur (fig. 6.5) resulting in material being transported into lower fields; possibly creating gulleys through lynchets. This higher energy inter-field erosion is more likely to become a component of valley fills.

6.4.2 Valley deposits

Sedimentation regimes in valley situations are more complex than those producing lynchets. Field boundaries will tend to receive unidirectional sedimentation, however valley sedimentation regimes are bi, if not multi-directional in origin, i.e. we must consider movement of material along the valley axis as well as immediately down the valley sides. Moreover, slope length may be greater in valley locations enabling higher energy deposits. This is borne out by the quantity of coarse components in some of the valley sections e.g., Kiln Combe, Itford Bottom, Chalton (Bell 1983) Ashcombe Bottom (Allen 1984a and b) and Devils Dyke (Ellis 1986). Although it has been suggested that the occurrence of lenses of large flint nodules may result from grazing animals kicking stones which roll into the valley bottoms, or by anthropogenic clearance of stones from fields, it is more likely that such horizons are the product of rill and gulley erosion. At Ashcombe Bottom a gravel fan consisting of wholly cortial flint nodules, mainly in the large and medium stone categories (according to Hodgson 1976) was encountered. Stone orientation was recorded. Preferential orientation was ca. 56° dip and ca. 90° strike which indicates the gravel fan is probably the result of a rill erosion event and compares well with many modern depositional fans (Boardman pers. comm.). The material creating the fan is a result of axial flow of sediment from a minor valley (fig. 6.6) and being deposited at the 'confluence' with Ashcombe Bottom. Minor rill erosion deposits were encountered in the archaeological deposits and equate well with modern (1982 and 1984) chalk rill fans at Ashcombe Bottom. This observation

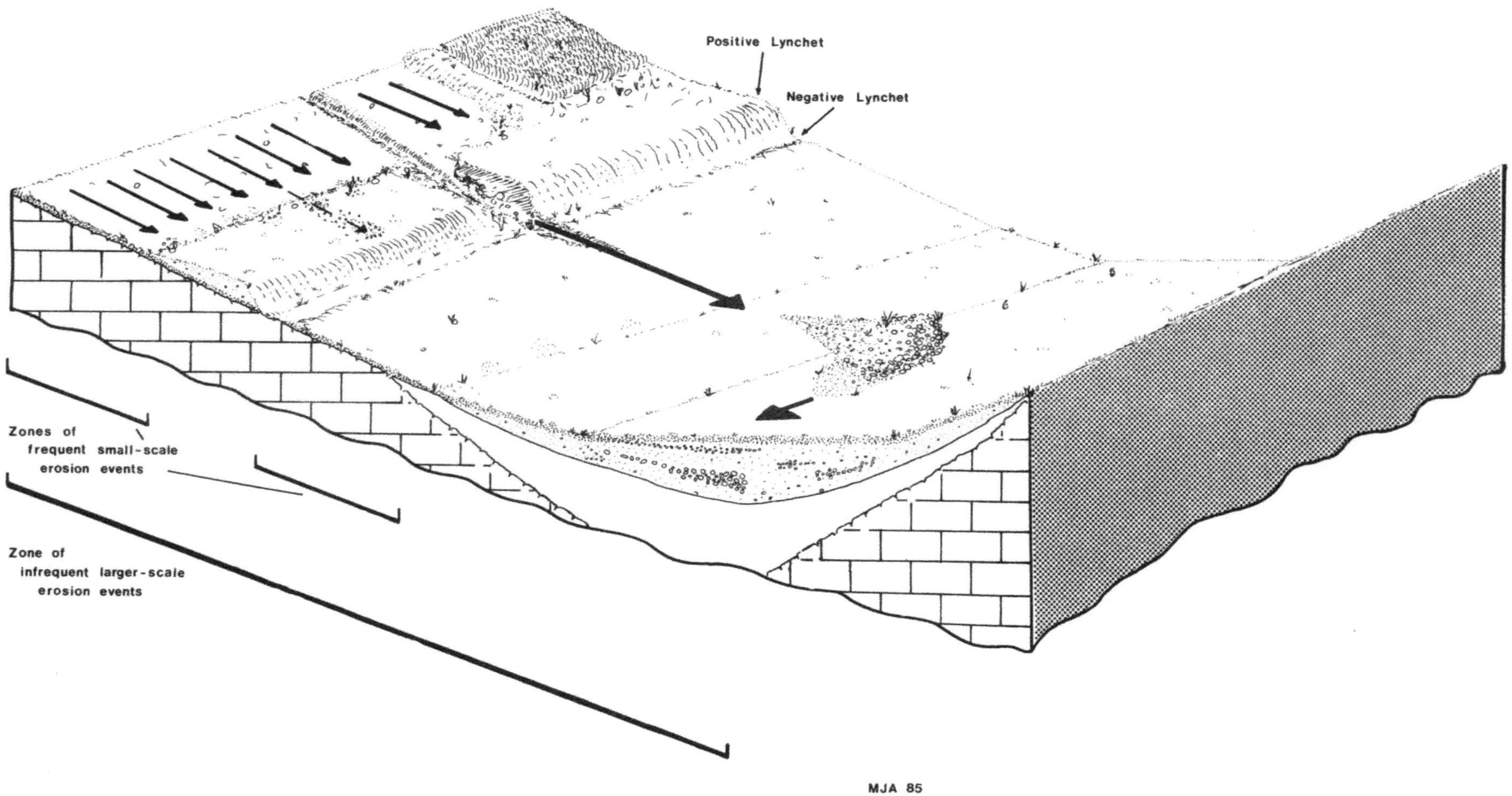

Fig. 6.5 Hypothetical chalkland landscape illustrating erosion regimes.

a

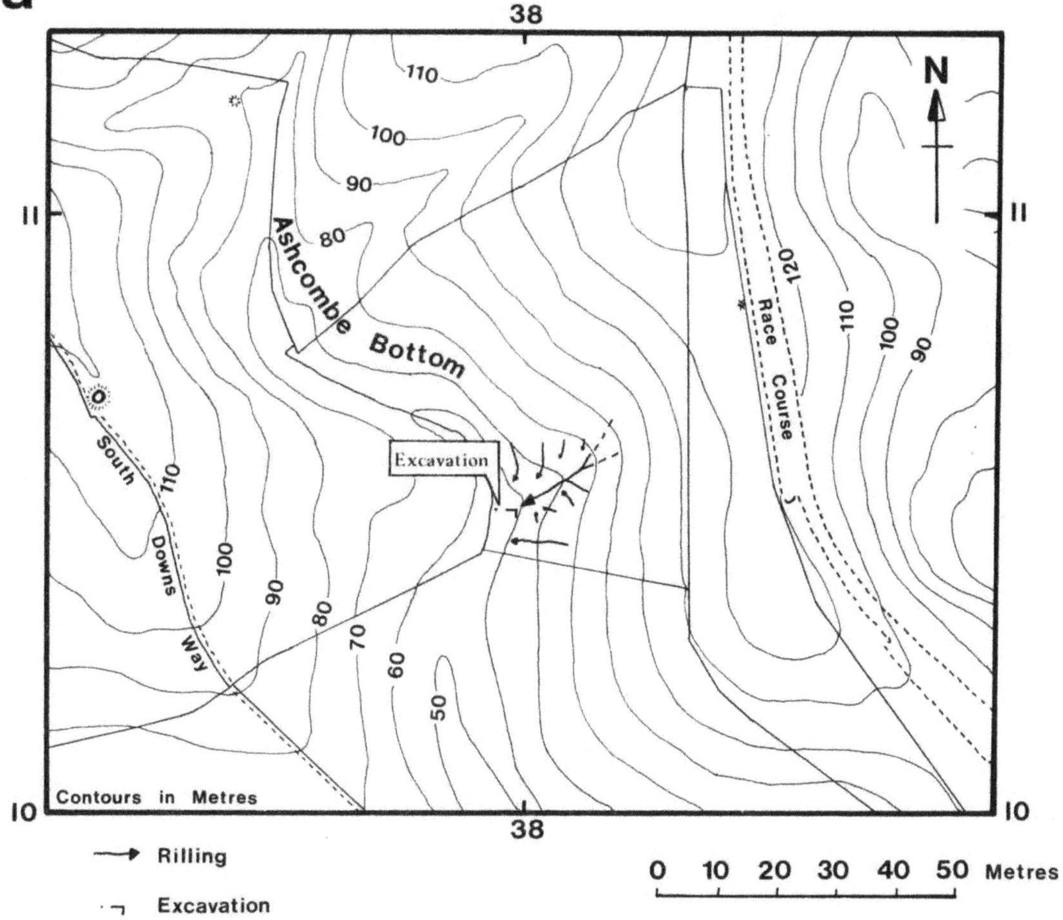

Fig. 6.6a Location of Ashcombe Bottom excavation showing rilling which occurred in 1984.

b

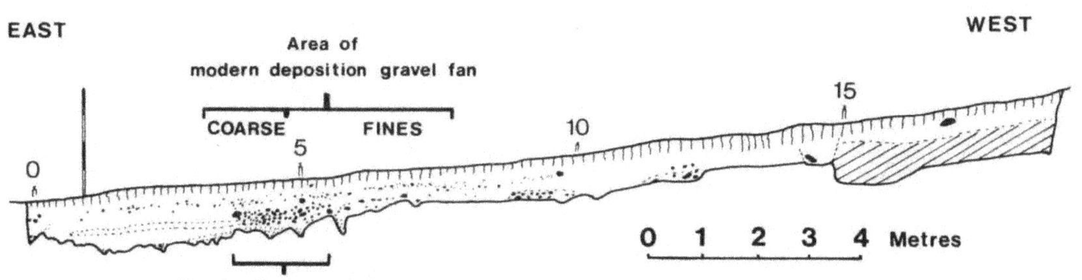

Fig. 6.6b Section of Ashcombe Bottom showing Bronze Age and modern depositional features.

probably accounts for the chalky lenses seen at Kiln Combe (layers 7 and 10) and illustrates that at both Ashcombe Bottom and Kiln Combe, high and low energy rill erosion occurred on the adjacent slopes, or up valley, in the past. Both are associated with Bronze Age horizons. Examination of these deposits in plan may allow the direction of depositional flow to be evaluated.

6.5 Modern erosion analogues

The interpretation of chalky and flint gravel horizons is only now made possible with the publication of observations of modern erosion analogues. Observations of Boardman and Stammers (1984) and Boardman and Robinson (1985) at Bevendean, Sussex indicate the scale and nature of rill erosion and its capacity to deposit numerous large flints as gravel fans and tons of fine sediments on valley floors.

At Chalton small scale rill erosion partially infilled with water and sediment an archaeological excavation which was located in the valley floor. More significant, however, was the erosion observed by the writer at Ashcombe Bottom in 1984. Excavation and investigation of the colluvial sediments in the valley included a Bronze Age gravel fan and a series of well dated chalky lenses, the latest situated on a post 1945 palaeosol and probably being the result of the onset of tillage in 1961 (C. Rae pers. comm.). After excavation had been completed and the trench backfilled, a storm on the night of Friday September 14th 1984 resulted in rilling of the adjacent field and produced a gravel fan and series of fans of chalky rill washes as well as depositing 17cm of fine sediments directly over the backfilled, buried archaeological sequence. Thus possible analogues of prehistoric erosion were fortuitously observed which demonstrated the range of material that may be deposited in a single erosive event (Plates 6.I and II).

If therefore, much of the colluviation we are dealing with in prehistory is a result of rilling, this has specific implications upon both the archaeological and environmental artefact removal. Boardman (1984) noted rills and gulleys scouring through the soil and into the chalk substrate. Therefore there is a possibility of gross mixing of artefacts and ecofacts from throughout the soil profile; i.e. material will not be uniformly stripped from the land surface and samples of the entire archaeological record may be washed into valley locations. However, evidence from the Bourne Valley and Ashcombe Bottom suggest that prehistoric erosion was primarily restricted to topsoil.

It may be that this 'fluvial' removal of material by large scale intermittent events allows the orientation of the magnetic domains which has been recorded in colluvium. This in turn provides the potential to date such accumulations by archaeomagnetic dating. However, it is still difficult to comprehend how such orientations remain fossilized with the abundant evidence of subsequent anthropogenic activity.

Much of the colluvium observed in valley and lynchet locations, although varied in depositional rate, can be attributed to anthropogenic disturbance and tillage. Modern erosion has been observed in Sussex occurring predominantly after harvest (i.e.

Plate 6.I Ashcombe Bottom during excavation, July 1984.
Photo: Mike Allen

Plate 6.II Ashcombe Bottom after erosion event, September 14th 1984.
Photo: Mike Allen

September) when the fields are dry and bare and therefore highly susceptible to run-off erosion caused by the first of the winter storms (Boardman and Stammers 1984; Boardman and Robinson 1985; Allen 1984b). Moreover, Boardman (pers. comm.) has shown that frequent larger scale erosion events may be considered a product of autumn sowing. Prehistoric man certainly had both the capacity (Reynolds 1985) and the crops (Jones 1981) to undertake autumn sowing. The potential for winter cereals in the pre-Roman period is hinted at by Jones (1981) and also by Reynolds (1981; 1985) who provides more empirical evidence for this regime. This evidence combined with both the modern observations of erosion and those recorded in the archaeological investigations discussed make this hypothesis an attractive one for many of the deposits. However, there are difficulties in making direct comparisons between modern and past episodes of erosion. Different methods of tillage and crop husbandry produce different microenvironments (Bell 1986).

6.6 Agriculture and past soil regimes

Although about 80% of the East Sussex chalk Downlands are under the plough today, in the 1930s only 20% was arable (Boardman and Stammers 1985). This change is predominantly due to the advent of artificial fertilisers and the greater tractive power of farm machinery making it possible to cultivate larger and steeper fields. Although modern technology is responsible for today's extensive arable system, it is evident that quite extensive farming was also practised in prehistory. The so-called 'Celtic' field systems were a ubiquitous feature of the Downland landscape until recent agricultural expansion removed much of this fossilized landscape. Holleyman's survey of 'Celtic' field systems in the Brighton district (1935) shows how extensive such systems were, and his areas mapped as "probable extent of celtic lynchets" is conservative in the author's opinion. Extensive agricultural systems were probably founded upon a quite different pedogenic regime.

From work on valley and lynchet sediments it is obvious that soil has been lost from the Downs. Although Evans (1975, 153) postulates that this does not necessarily require a reduction in soil depth, due to the natural pedogenic equilibrium, it seems likely that soil depth was greater in the past especially in view of the extensive farming practices in prehistory. Moreover, Evans (1975) suggests that due to the high $CaCO_3$ contents the earthworm fauna would be high allowing bioturbation and high organic levels, however Romans and Robertson (1983) have indicated a lack of earthworm activity in Neolithic soils and this compliments the hypothesis, discussed below, of thicker more acidic soils.

The thin grey and brown rendzinas that survive on the Downs today are classified as grade 3 soils (Bibby and Mackney 1969). This is in part due to their ubiquitously loamy nature (Wooldridge and Linton 1933) being characteristically silt-rich (60%) (Hodgson 1967, 131). This is probably the remnants of insoluble, non-calcareous silt-sized minerals resulting from loess deposits. Catt (1978) suggests an original deposit of between 1-4m of loess which extended across much of SE England but has now been grossly denuded by subsequent erosion events, many of which were prior to man's activity. However, Catt does attribute much deflation of the fertile loessic deposits to man. Soil depth was greater in prehistory and Bell's estimates of soil loss (Table 6.1) compare well with those calculated by Evans (1981) but these must be considered as minimum figures since much soil material may be flushed out of the valley 'reservoir' location. This hypothesis might help explain the lack of Neolithic woodland evidence in many colluvial sequences examined.

The greater depth and loessic nature of the prehistoric soils would have supported a different pedological regime; perhaps a calcareous brown earth or brown earth which would be more acidic and would support a different vegetational regime. This hypothesis lends credence to the *Pinus* charcoals recovered from a subsoil hollow at Itford dated to the Mesolithic BM-1544, 6820 ± 85 bc, and to the *Pinus* pollen that Thorley identified in the Vale of the Brooks existing into the middle Bronze Age (Thorley 1971; 1981) as pine growing on chalk tends to suffer from chlorosis, a condition arising from its inability to obtain sufficient magnesium in the presence of calcium carbonate. A thick fertile loessic soil would be very attractive to man but would be highly susceptible to aeolian erosion and slaking once tilled. Moffat (unpublished) provides a strong correlation between ancient field systems in Hampshire, and thin grey rendzinas, which in the author's opinion demonstrates the gross change in the landscape as a result of man and that the present Downland catena can be seen as a product of man's activity.

6.7 Environment in late Neolithic / early Bronze Age in south-east England

The evidence discussed briefly above from the sites at Kiln Combe, Itford Bottom, Chalton, Bishopstone, Bourne Valley and Ashcombe Bottom together with other palaeoenvironmental evidence will allow a more regional appraisal. Here I wish briefly to discuss the settlement and environment of the late Neolithic and early Bronze Age in SE England to allow a true integration of both archaeological and environmental data but also to provide a more specific regional data base than those hypothetical ones used before.

Evans' work (e.g. 1971) on the molluscan sequences at South Street long barrow has shown large scale deforestation and the onset of agriculture during the Neolithic in Wiltshire. This is reiterated by Waton's work at Winnal Moors, Hampshire (1982). For many years this hypothesis of Neolithic clearance has been extended to include much of SE England. Now however there is a body of palaeoenvironmental evidence to modify this hypothesis for the counties of Surrey, Kent and Sussex.

Thomas (1982) has recently shown from molluscan analysis that the five Sussex causewayed enclosures (Whitehawk, Offham, Bury Hill, Combe Hill and the Trundle) all show localised temporary deforestation of deciduous woodland which was allowed to regenerate rapidly afterwards. Moreover, the dry valley sequence at Brook (Rifle Butts) Kent, also produced molluscan evidence for localised Neolithic clearance followed by regeneration (Kerney et al. 1964, 165). Major permanent clearance in the south-east seems to have occurred in the early Bronze Age (Beaker period) or middle Bronze Age. Major clearance at Brook based on molluscan evidence from the Devils Kneadingtrough (Borehole V) is dated to the Beaker period (ca. 1700 BC). Corroborative evidence is provided by the pollen data from Wingham near Canterbury where the "consistently low AP/NAP ratio indicates that from as early as 1600 BC the region had been largely deforested" (Godwin 1962, 97). Macphail and Scaife (in press) have suggested that much of the erosion of the Surrey heaths was initiated in the Bronze Age and Allen (forthcoming) has demonstrated that primary woodland probably existed on the Hogs Back, Surrey, into the Bronze Age when it was locally removed for the construction of the Bronze Age Bell Barrow of Wen Barrow (Clark pers. comm.).

Sussex, however, provides a better environmental coverage both spatially and temporally. Thorley's work (1981) on the Vale of the Brooks in the Ouse Valley showed temporary vegetation clearance in the Neolithic (consistent with Thomas 1982), followed by more permanent clearance in the middle Bronze Age ca. 1240 $\pm$ 125 bc. Similarly Itford Bottom has produced a Beaker / early Bronze Age of 1770 $\pm$ 120 bc (BM-1545) for clearance, which was obtained from charcoals on a palaeosol which resulted from burning prior to cultivation (Bell 1983).

Cultural evidence from the colluvium at Kiln Combe (Bell 1982; 1983) and Ashcombe Bottom (Allen 1984a; 1984b) indicates late Beaker occupation sites buried well beneath the present valley floor. At both sites artefacts were associated with possible cultivation marks. The subsequent hillwash in both cases indicates prolonged open environments. We can conclude that in the south-east permanent clearances occurred in the Bronze Age with only minor clearances in the Neolithic.

The discovery of two Beaker occupation sites in five valley investigations is probably significant and has major implications for the prehistoric settlement pattern in Sussex and SE England. Champion (1982) has noted that in Kent the archaeological evidence for the Beaker period, other than funerary, is very slight. In Sussex Ellison (1978) showed 28 sites with funerary contexts, but only one potential settlement site, Belle Tout. There is also a lack of early Bronze Age settlement sites in Surrey. Within the three counties only three potential Beaker settlement sites have been located; the supposed type-site of Belle Tout (Bradley 1970), Kiln Combe (Bell 1983) and Ashcombe Bottom (Allen 1984a) all of which are situated in valley locations.

Belle Tout, which is enclosed within a valley entrenchment (Toms 1912), is buried under only a very thin veneer of colluvium. In a recent appraisal Bradley (1983) suggests that the site may have had an industrial rather than a domestic function. Kiln Combe produced a comprehensive array of domestic pottery (Bell 1983; Gibson 1982) from a 2m wide strip 10m long and concentrated at a depth of ca. 1.8m below the present valley floor. Furthermore Bell tentatively suggests the discovery of associated cultivation marks. Similarly at Ashcombe Bottom, just west of Lewes, the author recovered evidence of ard marks in a palaeosol which also contained both domestic Beaker pottery and typical early Bronze Age flint artefacts, predominantly horseshoe and discoidal scrapers. This site too was buried under ca. 80cm of colluvium. More recently, tentative evidence from Houndean Bottom (TQ

351 101) of several sherds of Beaker pottery in a valley location, which were brought to the surface by tree felling and rabbit activity, may provide another clue to further settlement. Further up the same valley system a previously undated valley entrenchment (Toms 1926; Allen and Fennemore 1984) has also produced decorated Beaker wares from a ditch exposure. The regular occurrence of material indicative of settlement buried under hillwash must not be purely coincidental. Five Sussex dry valley excavations have chanced upon two Beaker occupation sites where only one was known before. Furthermore the pottery from the Bourne Valley suggests the discovery of an Iron Age site buried beneath ca. 1.2m of colluvium. Even a medieval farmstead was blanketed by colluvium in Kiln Combe. All of these sites are unlocatable by normal archaeological reconnaissance. The excavations at Kiln Combe, Itford and Chalton were chosen in areas of known archaeological interest, whereas the Bourne Valley site was threatened by urban development and the location of the Ashcombe Bottom trench was predetermined by the farmer, Mr. C.J. Rae, due to problems of vehicular access. It is therefore evident that valley floor located settlements may be more common than previously accepted by archaeologists. The data from which past prehistoric distribution maps were produced is highly biased towards hilltop locations where material is readily exposed by erosion and biased against valley locations where the floors are constantly receiving sediment. Thus perhaps a re-evaluation of Beaker, if not prehistoric, settlement in SE England is needed in the light of the corpus of evidence from dry valley excavations. Present prehistoric settlement distribution patterns <u>must</u> at least be used with extreme caution as they probably show taphonomic factors of archaeological site preservation rather than the whole spectra of past activity areas.

6.8 Conclusions

In this paper I have attempted briefly to cover a large body of rapidly growing data and indicate the potential for both palaeoenvironmental and archaeological reconstruction from within and under hillwash. This study has shown that the Downlands are largely a result of human exploitation.

There is still much scope for improvement and some of the techniques discussed (e.g. magnetic susceptibility and archaeomagnetic dating) are still in their infancy regarding their application to these materials. The potential widespread archaeological implications of these investigations cannot be over stressed, let alone their combined potential for regional, local and micro-environmental determination.

6.9 Postscript

Since this paper was written three more Bronze Age sites have been discovered beneath colluvium; Middle Farm, Dorchester (middle Bronze Age), Strawberry Hill, Wiltshire (late Bronze Age) and Holywell Combe, Folkestone (Beaker). Further, work by Evans in the Kennet valley and R.W. Smith and R. Mercer at Irwene Meadow, Hambledon also recorded settlement evidence blanketed by sediments.

Acknowledgements

The topic of research here owes much to the work of, discussions with and encouragement given by, Dr. Martin Bell over the past five years, to whom I am eternally gratefull.

My own excavations at Bourne Valley and Ashcombe Bottom would not have been possible without the kind permission of Tescos Stores Ltd., and Messrs T.G. and C.J. Rae respectively. Financial support was provided by the Central Research Fund of London University, Harvey and Sons of Lewes, Sussex Archaeological Society, Mr. Chips of Lewes and A. Lewison Esq. Personnel and excavation equipment were provided by the Lewes Archaeological Group, Eastbourne Urban Medieval Excavation Project, Brighton and Hove Archaeological Society and the Association for Cultural Exchange. Many friends helped and in particular I wish to thank Messrs M. Brown, D. Gregory, P. Hill, R. Lloyd, R. Scott and A. Sayers and Ms. J. Crawford, L. Mount, J. Neate, L. Sherman, G. Whitman and L. Yaffe and John and Phoebe Williams. Special thanks are due to Lawrence Stevens, the director of the E.U.M.E.P., P. Stevens, P. Norman and C. Stevens all of whom undertook special responsibilities. Fruitful discussions and advice was readily given by R. Macphail, J. Boardman, P. Fisher and A. Moffat (soils and sediments), P. Drewett, R.W. Smith (archaeology), and A. Clark (magnetic properties). The Ancient Monuments Laboratory (HBMC) also provided facilities for archaeomagnetic dating and magnetic susceptibility measurement and the Soil Survey of England and Wales kindly provided a J.C.B. Finally I wish to thank J. Ede for her constant encouragement throughout this research, Miss R. Seager Smith for making many useful comments upon the script and to my parents for their endless support.

Many friends have provided useful comments throughout this research but all the opinions expressed are my own.

References

Aitken, M.J., 1974. Physics in archaeology. (2nd. ed.) Oxford: Clarendon Press.

Allen, M.J., 1983. Sediment analysis and archaeological data as evidence of the palaeoenvironments of early Eastbourne: the Bourne Valley excavation. London University, Institute of Archaeology, unpub. B.Sc., dissertation.

Allen, M.J., 1984a. Ashcombe Bottom excavations, first interim report. Lewes: Archaeological Group.

Allen, M.J., 1984b. Ashcombe Bottom excavations. Sussex Archaeological Society Newsletter 44.

Allen, M.J., forthcoming. Land mollusca from Wen Barrow; a Bronze Age bell barrow, Surrey. In: A.J. Clark Excavations of a Bronze Age barrow, Surrey.

Allen, M.J. and Fennemore, A.V., 1984. Field boundary ditch, Cuckoo Bottom, Lewes. Sussex Archaeological Collections 122, 207-208.

Allen, M.J. and Macphail, R.I., 1985. Review and discussion of magnetic susceptibility and micromorphology: examples of Hazelton Long-Cairn and Seaford Head Hillfort, Sussex. Unpub. Ancient Monuments Laboratory Report No. 4583.

Allen, M.J. and Macphail, R.I., 1987. Micromorphology and magnetic susceptibility: their combined role in interpreting archaeological soils and sediments. In: N. Fedoroff, L.M. Bresson and M.A. Courty (eds.), Soil micromorphology, 669-676. Paris: L'Association Français pour l'Etude du Sol.

Bell, M.G., 1977. Bishopstone excavations. Sussex Archaeological Collections 115.

Bell, M.G., 1981a. Valley sediments as evidence of prehistoric land-use in South East England. London University, Institute of Archaeology, Ph.D. thesis.

Bell, M.G., 1981b. Valley sediments and environmental change. In: M. Jones and G. Dimbleby (eds.), The environment of man: the Iron Age to the Anglo-Saxon period. BAR, Brit. Ser. 87, 75-91.

Bell, M.G., 1982. The effects of land-use and climate on valley sedimentation. In: A.F. Harding (ed.), Climatic change in later prehistory, 127-142. Edinburgh: University Press.

Bell, M.G., 1983. Valley sediments as evidence of prehistoric land-use on the South Downs. Proceedings of the Prehistoric Society 49, 118-150.

Bell, M.G., 1986. Archaeological evidence for the date, cause and extent of soil erosion on the chalk. South East Soils Discussion Group Journal 3, 72-83.

Bibby, J.S. and Mackney, D., 1969. Land Use Capability Classification. Harpenden: Soil Survey Technical Monograph 1.

Bintliff, J., 1977. Natural environment and human settlement in Prehistoric Greece. BAR, Intern. Ser. 28.

Boardman, J., 1984. A morphometric approach to soil erosion on agricultural land near Lewes, East Sussex. In: C.T. Lukehurst and R.L. Grant (eds.), Issues in Countryside Research. Brighton: Polytechnic, Kingston/Brighton Research Papers, 1-10.

Boardman, J. and Robinson, D.A., 1985. Soil erosion, climatic vagary and agricultural change on the Downs around Lewes and Brighton, autumn 1982. Applied Geography 5, 243-258.

Boardman, J. and Stammers, R.L., 1984. Soil erosion and flooding on Downland areas. The Surveyor 164, 8-11.

Bowen, H.C., 1961. Ancient Fields. London: British Association for the Advancement of Science.

Bradley, R., 1970. The excavation of a Beaker settlement at Belle Tout, East Sussex, England. Proceedings of the Prehistoric Society 36, 312-379.

Bradley, R., 1983. Belle Tout - Revision and re-assessment. In: P.L. Drewett, The archaeology of Bullock Down, Eastbourne, East Sussex: the development of a landscape, 12-20 Sussex Archaeological Society Monograph 1.

Catt, J.A., 1978. The contribution of loess to soils in lowland Britain. In: S. Limbrey and J.G. Evans (eds.), The effects of man on the landscape; the lowland zone. CBA Research Report 21, 12-20.

Champion, T.C., 1982. The Bronze Age in Kent. In: P.E. Leach (ed.), Archaeology in Kent to A.D. 1500. CBA Research Report 48, 31-39.

Clark, A.J., 1980. Magnetic dating, Sussex Archaeological Collections 118, 7-13.

Clark, A.J., 1983. The testimony of the topsoil. In: G.S. Maxwell (ed.), The impact of aerial reconnaissance on archaeology. CBA Research Report 49, 128-135.

Clarke, A.F., 1983. An excavation of a lynchet south west of Heathy Brow Iron Age settlement. In: P.L. Drewett, The archaeology of Bullock Down, Eastbourne, East Sussex: the development of a landscape, 17-21, Sussex Archaeological Society Monograph 1.

Dearing, J.A., Elner J.K. and Happey-Wood, C.M., 1981. Recent sediment influx and erosional processes in a Welsh upland lake-catchment based on magnetic susceptibility measurement. Quaternary Research 16, 356-372.

Dimbleby, G.W., 1976. Climate, soil and man. Philosophical Transcations of the Royal Society, London B 275 197-208.

Ellis, C., 1986. The post-glacial molluscan succession of the South Downs dry valleys. In: G. de G. Sieveking and M.B. Hart (eds.), The scientific study of flint and chert: proceedings of the fourth international flint symposium, 177-184. Cambridge: University Press.

Ellison, A., 1978. The Bronze Age in Sussex. In: P.L. Drewett (ed.), Archaeology in Sussex to A.D. 1500. CBA Research Report 29, 30-37.

Evans, J.G., 1966. Late glacial and post-glacial subaerial deposits at Pitstone, Bucks. Proceedings of Geologists Association 77, 347-364.

Evans, J.G., 1971. Habitat change on the calcareous soils of Britain: the impact of Neolithic man. In: D.D.A. Simpson (ed.), Economy and settlement in Neolithic and early Bronze Age Britain and Europe, 27-73. Leicester: University Press.

Evans, J.G., 1972. Land snails in archaeology. London: Seminar Press.

Evans, J.G., 1975. The environment of early man in the British Isles. London: Paul Elek.

Evans, J.G. and Valentine, K.W.G., 1974. Ecological change induced by prehistoric man at Pitstone, Bucks. Journal of Archaeological Science 1, 343-351.

Evans, R., 1981. Potential soil and crop losses by erosion. Paper given at conference on soil and crop loss - developments in erosion control. Stoneleigh 3.12.81.

Fisher, P.F. and Macphail, R.I., 1985. Studies of archeological soils and deposits by micromorphological techniques. In: N.R.J. Fieller, D.D. Gilbertson and N.G.A. Ralph (eds.), Palaeoenvironmental Investigations BAR, Intern. Ser. 258, 93-112.

Gibson, A.M., 1982. Beaker domestic sites; a study of domestic pottery of the third and early second millennia BC in the British Isles. BAR, Brit. Ser. 102.

Godwin, H., 1962. Vegetational history of the Kentish Chalk Downs as seen at Wingham and Frogholt, Veröffertlichungen des Geobotanisches Instituts der Techn. Hochschule Rübel Zürich, H37, 83-99.

Groenman-van Waateringe, W., 1979. Nogle aspekter af jernalderens agerbrug i Holland og NV Tyskland. In: H. Thrane (ed.), Fra jernalder til Middelalder Skrifder fra Historisk Institut, Odense Universitet 27, 75-86.

Hall, N.W., 1987. An application of micromorphology to evaluating the distribution and significance of erosion by water. In: N. Fedoroff, L.M. Bresson and M.A. Courty (eds.), Soil micromorphology, 437-443. Paris: L'Association Français pour l'Etude du Sol.

Hodgson, J.M., 1967. Soils of the West Sussex coastal plain. Bulletin of the Soil Survey of Great Britain 3.

Hodgson, J.M., 1976. Soil Survey field handbook. Harpenden: Soil Survey Technical Monograph 5.

Holleyman, G.A., 1935. The Celtic field-system in South Britain: a survey of the Brighton district. Antiquity 9, 443-454.

Jones, M., 1981. The development of crop husbandry. In: M. Jones and G. Dimbleby (eds.), The environment of man: the Iron Age to the Anglo-Saxon period. BAR Brit. Ser. 87, 95-127.

Kerney, M.P., Brown, E.H. and Chandler, T.J., 1964. The late glacial and post-glacial history of the chalk escarpment near Brook, Kent, Phil. Trans. Roy. Soc. London B 248, 135-204.

Kwaad, F.J.P.M. and Mücher, H.J., 1977. The evolution of soils and slope deposits in the Luxembourg Ardennes near Wiltz, Geoderma 17, 1-37.

Kwaad, F.J.P.M. and Mücher, H.J., 1979. The formation and evolution of colluvium on arable land in northern Luxembourg, Geoderma 22, 173-192.

Longworth, G. and Tite, M.S., 1977. Mössbauer and magnetic susceptibility studies of iron oxides in soil from archaeological sites, Archaeometry 19, 3-14.

Macphail, R.I., 1983. Micromorphological report on Chalton B layer 3. In: M.G. Bell, Valley sediments as evidence of prehistoric land-use on the South Downs, Proceedings of the Prehistoric Society 49, microfiche 55-57.

Macphail, R.I., Romans, J.C.C. and Robertson, L., 1987. The application of micromorphology to the understanding of Holocene soil development in the British Isles; with special reference to Holocene cultivation in the British Isles. In: N. Fedoroff, L.M. Bresson and M.A. Courty (eds.), Soil micromorphology, 647-656. Paris: L'Association Français pour l'Etude du Sol.

Macphail, R.I. and Scaife, R.G., in press. The geographical and environmental background. In: J. Bird and D.G. Bird (eds.), The Archaeology of Surrey, Palaeolithic to 1540 A.D. Surrey Archaeological Society Monograph.

Moffat, A.J., The distribution of "Celtic fields" on the east Hampshire chalklands. Unpub. MS.

Mücher, H.J., 1974. Micromorphology of slope deposits: the necessity of classification. In: G.K. Rutherford (ed.), Soil microscopy, Proceedings of the 4th International Working Meeting on Soil Micromorphology. Ontario, 553-556.

Mullins, C.E., 1974. The magnetic properties of the soil and their application to archaeological prospecting, Archaeo-Physica 5, 180-315.

Oldfield, F., Dearing, J.A., Thompson, R. and Garret-Jones, S., 1978. Some magnetic properties of lake sediments and their possible links with erosion rates, Polskie Archiwum hydrobiolgii 25, 321-331.

Potter, T.W., 1976. Valleys and settlement: some new evidence, World Archaeology 8, 207-219.

Reynolds, P.J., 1981. New approaches to familiar problems. In: M. Jones and G. Dimbleby (eds.), The environment of man: the Iron Age to Anglo-Saxon period. BAR, Brit. Ser. 87, 19-49.

Reynolds, P.J., 1985. Iron Age agriculture reviewed. Wessex Lecture I, CBA Archaeology Group 12.

Romans, J.C.C. and Robertson, L., 1983. The general effect of early agriculture on the soil. In: G.S. Maxwell (ed.), The impact of aerial reconnaissance on archaeology. CBA Research Report 49, 136-141.

Tarling, D.H., 1971. Principle and applications of palaeomagnetism. London: Chapman and Hall.

Tarling, D.H., 1975. Archaeomagnetism: the dating of archaeological materials by their magnetic properties, World Archaeology 7, 185-197.

Tite, M.S. and Mullins, C.E., 1970. Electromagnetic prospecting on archaeological sites using a soil conductivity meter, Archaeometry 12, 97-104.

Tite, M.S. and Mullins, C.E., 1971. Enhancement of magnetic susceptibility of soils on archaeological sites, Archaeometry 13, 209-219.

Thomas, K.D., 1982. Neolithic enclosures and woodland habitats on the South Downs in Sussex, England. In: M.G. Bell and S. Limbrey (eds.), Archaeological Aspects of Woodland Ecology. BAR, Intern. Ser. 146, 147-170.

Thompson, R., Battarbee, R.W., O'Sullivan, P.E. and Oldfield, F., 1975. Magnetic susceptibility of lake sediments, Limnology and Oceanography 26, 687-698.

Thorley, A., 1971. Vegetational history in the Vale of the Brooks. In: R.B.G. Williams (ed.), Guide to Sussex excursions. Falmer: University of Sussex, Institute of British Geographers Conference, 47-50.

Thorley, A., 1981. Pollen analytical evidence relating to the vegetational history of the chalk, Journal of Biogeography 8, 93-106.

Toms, H.S., 1912. Excavations of the Beltout valley entrenchments, Sussex Archaeological Collections 55, 41-55.

Toms, H.S., 1926. Valley entrenchments east of the Ditchling Road, Brighton and Hove Archaeologist 3, 42-61.

Waton, P.V., 1982. Man's impact on the chalklands: some new pollen evidence. In: M.G. Bell and S. Limbrey (eds.), Archaeological aspects of woodland ecology. BAR, Intern. Ser. 146, 75-91.

Wooldridge, S.W. and Linton, D.L., 1933. The loam-terrains of South-east England and their relation to its early history, Antiquity 7, 297-310.

7. THE ARCHAEOLOGICAL COMPONENT OF AGRICULTURAL SOILS IN THE MIDDLE EAST: THE EFFECTS OF MANURING IN ANTIQUITY

T.J. Wilkinson

Abstract

This paper describes first the distribution of extensive scatters of artefacts around four sites or site complexes in the Middle East. There follows a summary of the distribution of such artefacts through the soil profile. In turn, these scatters can be used to show how the associated soils have been altered by ploughing, deflation, sedimentation and erosion since the arrival of the artefacts. Finally, the relationship between artefact scatters and soil phosphates is shown for one of the case study areas. The scatters are interpreted as resulting from the use of settlement-derived manure and compost on fields adjacent to the settlements.

7.1 Introduction

The presence of artefacts and other archaeological debris within agricultural soils has long been recognized, but to date few comprehensive studies have been made to place such occurrences within an agrarian or edaphic framework. In this paper, the evidence for extensive artefact scatters within four areas of the Middle East is described, and related to the local settlement archaeology and soil profile development. This is followed by a case study from the Turkish survey area which attempts to relate soil phosphate values to the history of soils and land use of the area. Finally, a brief and necessarily preliminary review of the evidence presented, will attempt to interpret artefact scatters in the light of ancient manuring practice, environmental, and demographic history.

The first systematic attempt to relate extensive artefact scatters to early agricultural practice was by Rhodes in his studies of later prehistoric field systems of the Berkshire Downs (Rhodes 1950; Fowler 1981, 167). Since that time, artefact scatters of various dates have been recorded from Britain (Williamson 1984; Gaffney and Tingle 1985, 71; and P. Hayes, pers. comm.); Iran, Oman, Syria, Turkey (Wilkinson 1982 and below); Tunisia (Greene 1984) and Yemen (Whitcomb forthcoming). However, probably the most comprehensive studies of early anthropogenically amended soils have been made in the Netherlands, where plaggen soils enriched with humic soil material, animal manure and sometimes organic settlement refuse, have been subjected to detailed study (van de Westeringh, Groenman-van Waateringe, Heidinga and Bakels, this volume). Most of the above references interpret the presence of artefacts within agricultural soils as a result of the spreading of settlement-derived manure, street sweepings, "night soils" from cess pits etc. Inevitably, such sources include artefacts which eventually become incorporated into the soil profile far from their point of original use. Such an explanation may therefore account for the presence of archaeological debris within soils of ancient agricultural oases. As Gorbunov has pointed out within the context of Soviet central Asia:

> "Irrigation sedimentation over several centuries or even millennia results in an increment of oasis soils with the formation of a new irrigational horizon. The accumulation of such a horizon was enhanced in the past by the widespread method of adding earthy material upon the fields: material taken from canal deposits, earth on hillocks, old mud brick constructions, etc. Man-made horizons in soil of such ancient oases as those of Khwarizm, Bukhara, Samarkand, Murgab, reach a thickness of 1-2m or more." and: "Irrigational horizons are characterized by various archaeological inclusions: potsherds, small fragments of burned brick and of charcoal, bones - which help to distinguish such horizons when studying the profile of oasis soils in the field."
> (Gorbunov 1965, 39; see also Subramanian and Zarif-Salem 1965, 69).

The practice of manuring with urban-derived organic wastes is documented in a wide range of literature (summarised in Wilkinson 1982). For example, Chinese cities produced larger quantities of human excrement than the surrounding villages. This waste was removed from the cities, composted with vegetable matter and used as manure. Hence Chinese cities were girdled with concentric rings of green, the densest being near the city, and the faintest being further out as a result of the increased cost of manure away from the City (Lattimore 1979, 36). Similar processes have been observed in the recent past in eastern Iran where field areas around larger settlements are better fertilised than those around smaller villages (English 1966, 122). Furthermore, classical writers have noted the spreading of urban-derived manure around cities in the Eastern Mediterranean (Semple 1932).

As well as enriching the soil in phosphates, potash, nitrogen and various trace elements, the application of manures and composts enhances the physical properties of the soil and increases its water retention potential. Therefore although manuring is most effective in the Middle East under an irrigation regime, by enhancing the moisture retention properties of the soil, composts could be effective within a regime of dry farming (Wilkinson 1982, 324).

7.1.1 The case study areas

The following site territories are all located in the Middle East in areas sufficiently unencumbered by modern settlement for site survey to have been pursued unimpeded over wide areas. Following a brief summary of the geography of each area, the distribution of artefacts over the terrain is described. Then the distribution of artefacts through the soil profile is examined and interpreted in terms of early land use practice and post-cultivation processes of sedimentation of soil formation.

Siraf

Located on the Iranian side of the Persian Gulf where the Zagros fold mountains meet the coast (fig. 7.1). Excavation of the Sasanian-early Islamic city was directed by Dr. D. Whitehouse under the sponsorship of the British Institute of Persian Studies. Mean annual rainfall is approximately 200mm per annum and although dry farming yields results in most years, irrigation is used where conditions are suitable. Cultivable soils are limited to a narrow, stony coastal plain and small, cultivable enclaves within the often highly eroded mountains. Soil horizons are usually weakly developed or absent and the soils can usually be classed as calcareous, gypsic or locally saline regosols (F.A.O. class: Calcic Regosols, F.A.O. 1977, map 7.1).

Sohar

Located on the Batina coast of northern Oman (fig. 7.1). The early Islamic city reached its peak in prosperity during the 9th/10th centuries AD at approximately the same time as Siraf. The field and water supply systems were studied between 1975 and 1977 by the author as part of a programme of research and rescue archaeology within greater Sohar, sponsored by the Ministry of National Heritage and Culture, Muscat (Williamson 1973; Wilkinson 1975; 76; 77; Costa 1978). Estimated mean annual rainfall of 80mm has an extremely wide inter-annual variability and irrigation is essential for cultivated crops. The city and its surrounding palm gardens developed on a broad coastal plain built of sand bars, sabkha (salt flats) and infilled lagoons. Inland, fields had once spread over gently sloping accumulation plains and gravel fans which washed down from the northern Hagar mountains. The arid soils are saline near the coast, have low organic matter and available phosphate contents (ILACO 1975, v2) and have negligible development of pedogenic horizons. The soils of the Batina fall within the F.A.O. soil zone of Haplic Yermosols (F.A.O. 1977, map 7.1).

Tell Sweyhat

Situated on a left bank terrace of the River Euphrates near the western edge of the Syrian Jazirah (fig. 1). The early Bronze Age and Hellenistic site was excavated by a team directed by Dr. T. Holland and sponsored by the Ashmolean Museum, Oxford. The surrounding landscape was investigated by the author in autumn 1974 (Wilkinson in Holland 1976, 67-70). A mean annual rainfall of 200-300mm places the site close to the southern limit of dry-land farming but cereal crops can be produced in most years. The red-brown loamy soils which cover the Euphrates terraces often include indurated calcium carbonate subsoil horizons (Calcic Xerosols with a petrocalcic horizon, F.A.O. 1977, map 7.1).

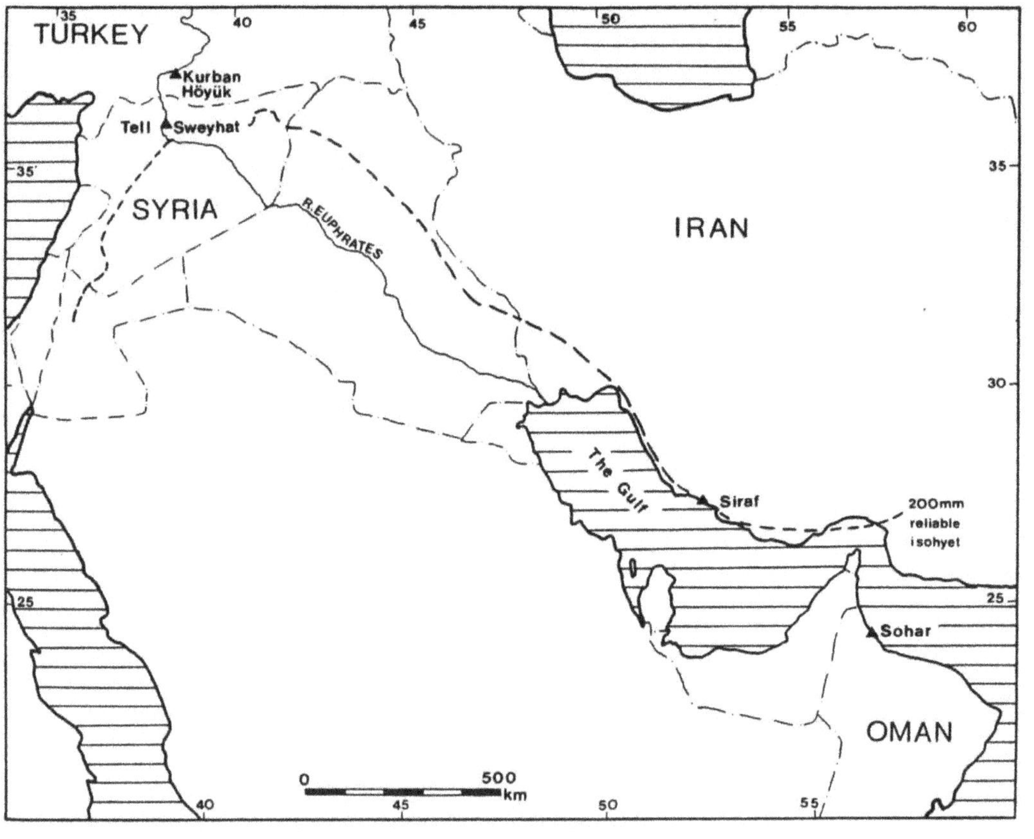

Fig. 7.1 Location of the four case study areas within the Middle East. Drawing: T.J. Wilkinson.

Kurban Höyük

Located 150km north of Tell Sweyhat on a left bank terrace of the Euphrates within SE Turkey (fig. 7.1). The chalcolithic, early Bronze Age and early Islamic site was excavated in conjunction with a detailed survey of its hinterland, as part of a five year programme (1980-84) directed by Dr. L. Marfoe of the Oriental Institute, University of Chicago and sponsored by the same Institute. The region, which morphologically is an extension of the Syrian Jazirah, is fringed to the north by the high Anti-Taurus mountains. A mean annual rainfall of 400-500mm allows the cultivation of cereals either every year or within an alternating fallow regime. The red-brown loam soils developed on the Euphrates terraces have weakly developed calcium carbonate B and C horizons and like the soils surrounding Tell Sweyhat, they can be classed as Calcic Xerosols (F.A.O. 1977, map 7.1).

7.2 Spatial distribution of artefacts

Except within the territory of Siraf, where field systems alone were mapped, surface sherd scatter densities were defined by means of the systematic collection of all artefacts from within sample squares laid out between 100 and 500m apart. Around Tell Sweyhat and Kurban Höyük, sample squares measured 10 x 10m, whereas behind Sohar where scatter densities were higher, sample squares were reduced to 5 x 5m or, in extreme cases to 2 x 2.

At Siraf, the presence of abandoned field systems defined by stone perimeter or terrace walls, enabled the approximate limits of early Islamic fields to be mapped (fig. 7.2). Within 50 minutes walk of the city, a scatter of early Islamic pottery, which included imported luxury wares, littered the surfaces of fields. Beyond this point, field surfaces were devoid of artefacts. The sherd scatters did not appear to have emanated from nearby rural buildings. Instead their overall distribution, and the inclusion of abundant luxury wares characteristic of the rich merchants' houses of the city, suggests that they came from the city, probably within organic refuse used as manure. Fields beyond 50 minutes walk of the city were evidently beyond the limit of economic haulage and were either manured from local sources or remained un-manured. Between 50 and 75% of the total field area was irrigated during early Islamic times (Wilkinson 1974, 129), and this practice, combined with manuring, must have ensured a reliable annual crop of vegetables, fruit, and possibly cereals. Nevertheless, it is unlikely that such a small cultivated area (691 ha) can have sustained the population of such a large city (Table 7.1).

In contrast, around the equally affluent trading emporium of Sohar, field systems were not defined by field walls. Instead, extensive scatters of potsherds, glass, brick fragments, slag, and marine shells spread over some 673 ha. These scatters coincided with the termini of numerous early Islamic open or underground water channels and wells. This suggests that, as at Siraf, scatters may have resulted from manuring with urban-derived organic materials (Wilkinson 1975; 1976). They appeared to define the perimeter of a complex of ancient cultivation which included closer to the coast, modern as well

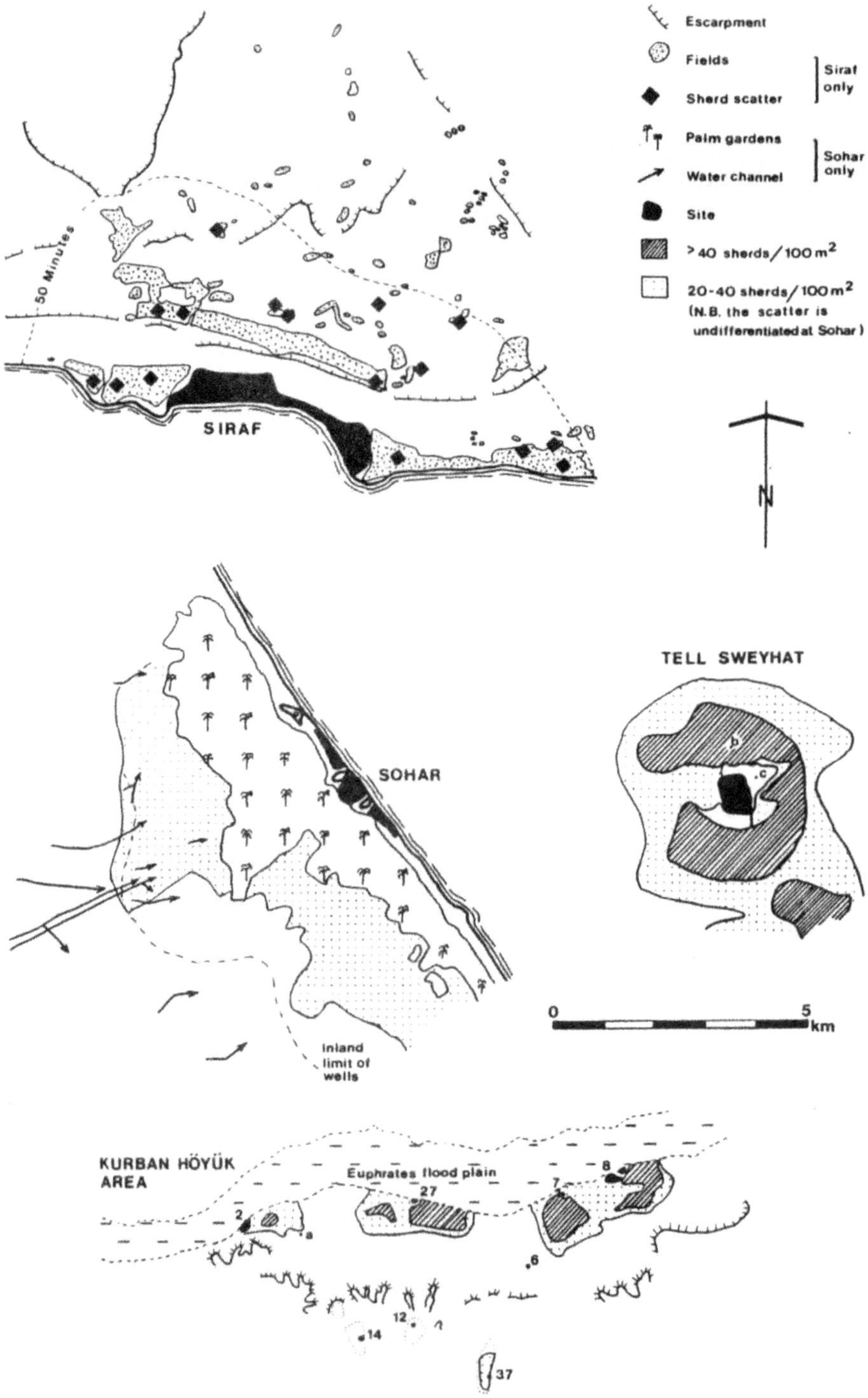

Fig. 7.2 Areas of inferred ancient cultivation within the case study areas (all to the same scale). _a_ on the Kurban Höyük map and _b_, _c_ on the Tell Sweyhat map indicate locations of soil profiles illustrated on fig. 3. Additional sites mentioned in the text are numbered on the Kurban Höyük map. Drawing: T.J. Wilkinson.

Table 7.1. Site area and calculated area of intensive cultivation for sites discussed in text

Site	Site area (ha)	Field area (ha)	Sherd scatter area (ha)	Cultivated area / Site area
Siraf	129.0	691.0	-	5.4
Sohar	73.0	-	3947.0	54.1
Tell Sweyhat	41.0	-	2608.0	63.6
Kurban Höyük area:				
Site 2	4.7	-	87.3	18.6
Site 27	0.8	-	236.5	315.3
Site 7	4.0	-	179.3	44.5
Site 8	6.5	-	152.0	23.4
Site 12	0.9	-	21.5	23.9
Site 14	1.2	-	12.8	10.6
Site 37	1.5	-	21.3	14.2

as probably early Islamic palm gardens (fig. 7.2). The total area of approximately 4000 ha of former cultivation was, almost certainly capable of supplying both the resident population and visiting ships.

In the Kurban Höyük survey area, artefact scatters were spread over terrain which surrounded a well-developed network of rural settlements. These attained their peak population during late Roman-early Byzantine times (4th-6th centuries AD). The settlements in question were mere villages, ranging in size from 0.9 to 6.5 ha, and the scatters, defined by 264 10 x 10m sample squares, were proportionally smaller than those around the afore-mentioned early Islamic cities. For example, the radii of sherd scatters ranged from 0.5 - 1.7km around the larger settlements of the main Terrace III, whereas smaller scatters of 0.25 - 0.7km radius surrounded the hamlets on the less well-watered Terrace I (figs. 7.2, 4 and 7 and Table 7.1).

Around Tell Sweyhat, artefact scatters were of similar density to those which surrounded the sites in Turkey, but they were of late 3rd millennium date. This corresponded to the maximum extent of the walled early Bronze Age town (Table 7.1).

With the exception of Siraf, the area of "field scatters" increased roughly in proportion to the area of the central site (Table 7.1). It is well known that principles of conservation of effort constrain the distribution of manure and other agricultural inputs to within a few kilometres of the central settlement (Chisholm 1979, 61). In the cases cited, the artefact distributions would appear to reflect such constraints but, the threshold distance beyond which manure was hauled increased with the size of the site and its populations. This was probably a function of both demand and supply, there being a greater demand for food as well as an increased supply of potential manure from the larger settlements (Lattimore 1979, 36).

Only Siraf, situated within the area of lowest agricultural potential, had an inferred cultivated area of less than might be expected for a city of its size. In addition to noting the aridity of the hinterland and its low agricultural potential, the 10th century writer Istakhri also emphasised the opulence of the city and its merchants (reviewed in Whitehouse 1968). Therefore, combining the historical and field evidence, it can be inferred that Siraf was probably unable to sustain its population by means of locally produced food, but was sufficiently affluent to import food from elsewhere in mainland Iran and from other communities within the Gulf.

Plant remains recovered from excavated contexts at Siraf (Williams 1973), Kurban Höyük (Miller in Marfoe 1986) and Tell Sweyhat, together with limited historical sources, provided some information of food consumption and crops grown. These appeared to have been similar to those of today with the southern cities of Siraf and Sohar relying upon irrigated dates, citrus fruits, cereals and lentils. In contrast, the two northern areas were part of a more extensive cereal and lentil economy, which was supplemented around Kurban Höyük by viticulture.

7.3 Artefact distributions within the soil profile

Within the context of man-made soils, such extensive artefact scatters can be seen to be only the most conspicuous skeletal part of a more substantial organic contribution made, in some cases over several millennia. The following examples illustrate how artefact distributions within the soil profile can vary depending upon the history of sedimentation and pedogenesis.

In the simplest case, the semi-arid zones around Kurban Höyük and Tell Sweyhat, it appears that artefacts became mixed throughout the soil A and upper B horizons. As a result, only the surface component of a three dimensional distribution is recorded by sample survey. For example, to the south of Kurban Höyük, where three soil pits were dug through red-brown loamy field soils, a surface scatter density of 10-20 sherds per $100m^2$ (10 x 10m square) was reflected by a gradually diminishing sequence down the soil profile (fig. 7.3a). Although most sherds were present within the soil Ap horizon, very small sherds were recorded to depths of 1.20m within the lower B horizon as a result of movement down cracks which separated soil structural units. A similar distribution was recorded at Tell Sweyhat (fig. 7.3b).

Evidently rather sparse surface sherd scatters ranging from 10 to

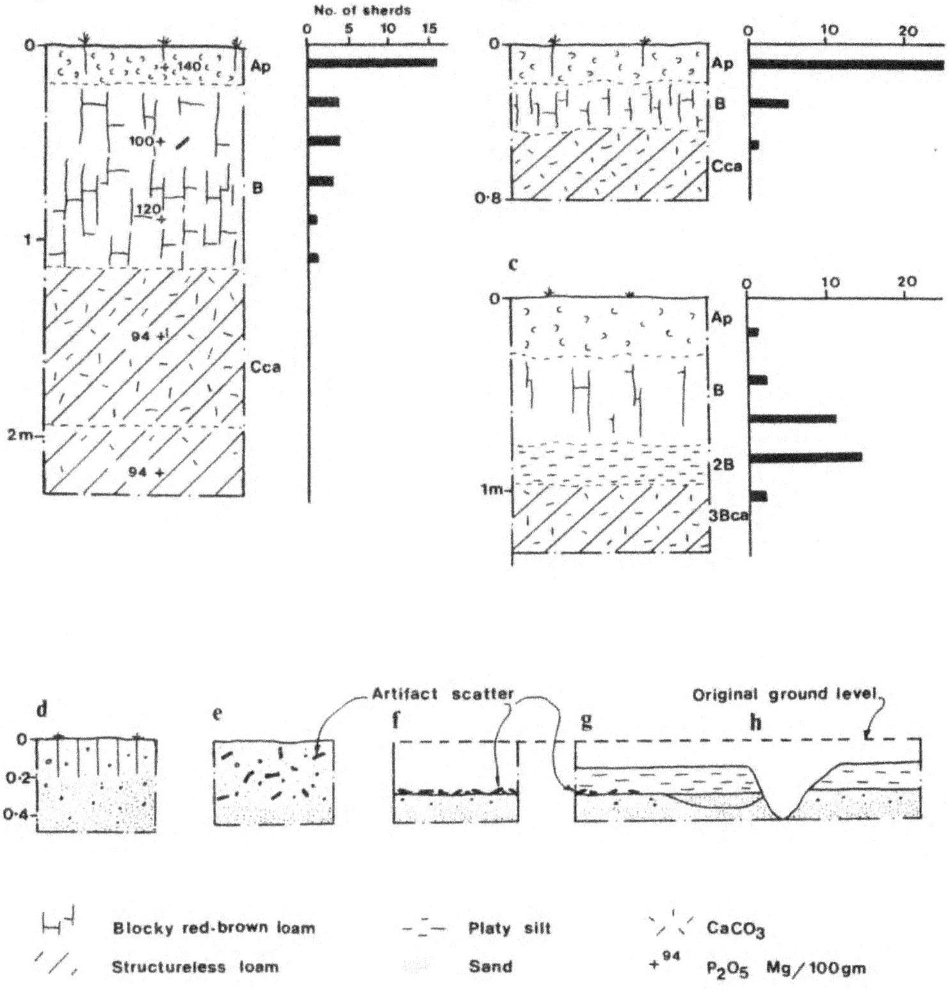

Fig. 7.3 Selected soil profiles from Kurban Höyük (a), Tell Sweyhat (b and c) and Sohar area (d-h), all to the same scale. Trenches a-c measured 1 x 1.5m in plan. Drawing: T.J. Wilkinson.

60 sherds per $100m^2$ are equivalent to much higher densities within the upper solum. For example, at Kurban Höyük, a surface scatter of 15 sherds per $100m^2$ was equivalent to 16 sherds per cubic metre of soil and near Tell Sweyhat, a surface scatter of 72 sherds per $100m^2$ was equivalent to 34 sherds per m^3 of soil.

The presence of an approximately planar distribution of artefacts across extensive tracts of land provides a useful datum for the measurement of localised aggradation and degradation. For example, within the roughly concentric configuration of sherd scatters around Tell Sweyhat, a lobate area was virtually devoid of artefacts. A soil pit excavated within this lobe intercepted a sherd-enriched horizon at a depth of 60-90cm; that is at the base of horizon B and within the top of horizon 2B (fig. 7.3c). The weak platy-structure loam (2B) overlay a palaeosol Bca horizon (3Bca) containing calcium carbonate nodules. Horizon 2B was, in turn overlaid by a loam with negligible horizon development, weak soil structure and incipient calcium carbonate accumulation (B). The aggraded sediment (2B) initially accumulated within a tranquil environment of deposition which was created when the later third millennium BC town wall was constructed and effectively dammed pre-existing drainage channels. Manuring with urban refuse, indicated on fig. 7.2 by surface artefact scatters, started during this period of the settlement's heyday. The contained artefacts then became mixed by ploughing through the lower part of horizon B and the upper part of 2B. Subsequently, aggradation since the early Bronze Age has resulted in the accumulation of the overlying virtually sherd-free loam. Finally, it can be concluded that the development of a weak sub-angular blocky structure and incipient calcium carbonate accumulation must have taken place over a period of some 4000 years.

As one moves towards more arid regions, deflation of agricultural topsoils appears to become more marked. For example, within the agricultural hinterland of Sohar, artefact densities normally ranged from 10-75 sherds per 2 x 2m sample square but occasionally attained peaks of 120-210 sherds (i.e. 2-18 and 30-53 sherds per m^2). Such high scatter densities were only limited to a surface 'lag' deposit which veneered an undisturbed B or C horizon. Only where soil horizons were buried beneath gravel mounds of upcast from early Islamic well mounds did intact cultivated horizons survive. In spite of their superficial resemblance to artefact scatters on habitation sites, the massive extent of these scatters (670 ha behind Sohar) as well as their association with irrigation systems, demonstrates them to have been part of an agricultural not an urban landscape. The surface lag deposit of artefacts has, of course, been allowed to survive because of the abandonment of the field areas. With renewed cultivation, scatters become re-incorporated into a cultivated Ap horizon and the evidence of deflation is lessened. This is, in part, the reason why sherd scatters were less prominent in the Turkish and Syrian study areas. If, however, the entire ploughsoil horizon had been deflated over the soil pits at Kurban Höyük (fig. 7.3a) and Tell Sweyhat (fig. 7.3b), total surface artefact densities of 10-20 sherds per m^2 would result. This falls within the middle range of values recorded from the Sohar field area.

Within the latter area, deposition of outwash silts from gravel fans as well as localised erosion, complicated the pattern described above. The schematic soil profiles shown on fig. 7.3d-h illustrate the development of an agricultural soil containing artefacts (3e) from a pre-existing weak horizoned profile (3d). With the deflation of the topsoil, a lag artefact horizon results (3f). Silt plain deposition from shallow runnels results in localised disruption of this horizon and its eventual burial. The low infiltration capacity of the resultant plain surface promotes storm runoff thus encouraging the development of gullies which further dissect the already highly fragmented agricultural horizon. All of the above processes were represented within the agricultural hinterland of Sohar and it is likely that such wholesale burial and removal of artefact-enriched agricultural soils may have resulted in the loss of some 480 ha of former cultivated land (Costa and Wilkinson, forthcoming).

7.4 Phosphate survey around Kurban Höyük

The sustained use of land over some 8000 years within the territory of Kurban Höyük provided a valuable opportunity to study the effects of extended periods of agriculture on the phosphate content of field soils. During the extensive sherd sampling programme, 219 soil samples were taken to determine the distribution of phosphate values spatially across the land surface and through selected soil profiles. It was hoped that, combined with artefact scatter data, phosphate analysis would enable areas of settlement-derived manuring to be distinguished as well as areas of less specific phosphate enrichment.

Phosphate analysis was conducted by D. Gurney on soil samples transported from Turkey (Gurney, in Wilkinson forthcoming). The method of analysis employed was based on the molybdenum blue method of Murphy and Riley (1962), modified for use in archaeological field survey by Hughes and Craddock (Sieveking et al., 1973, 195-196; Hughes 1976, 30-31). This method, which measures the inorganic component of the soil's total phosphate content, is generally accepted as being appropriate for the determination of ancient phosphates. Phosphates of anthropogenic origin are likely to have undergone conversion to the inorganic form and the effects of modern fertilizers are minimised (Hamond 1983, 62.).

Soil samples were collected from the following contexts within the detailed mapping area located within fig. 7.4 (i.e. the northern part of the area delimited for geomorphological mapping).

a) Thin soils developed on Eocene limestone (Soil Type 1, fig. 7.4).

b) Deep reddish-brown loams (Calcic Xerosols of pH 7-8) on Terrace I remote from archaeological sites (Soil Type 2).

c) Soils as for b), but from ploughsoils within areas of "field scatter" (Soil Type 2).

d) Soils as for c) but on Terrace II (Soil Type 2).

e) Six samples in vertical sequence from a soil profile exposed in Type 2 soils on Terrace III south of Kurban Höyük (E fig. 7.4).

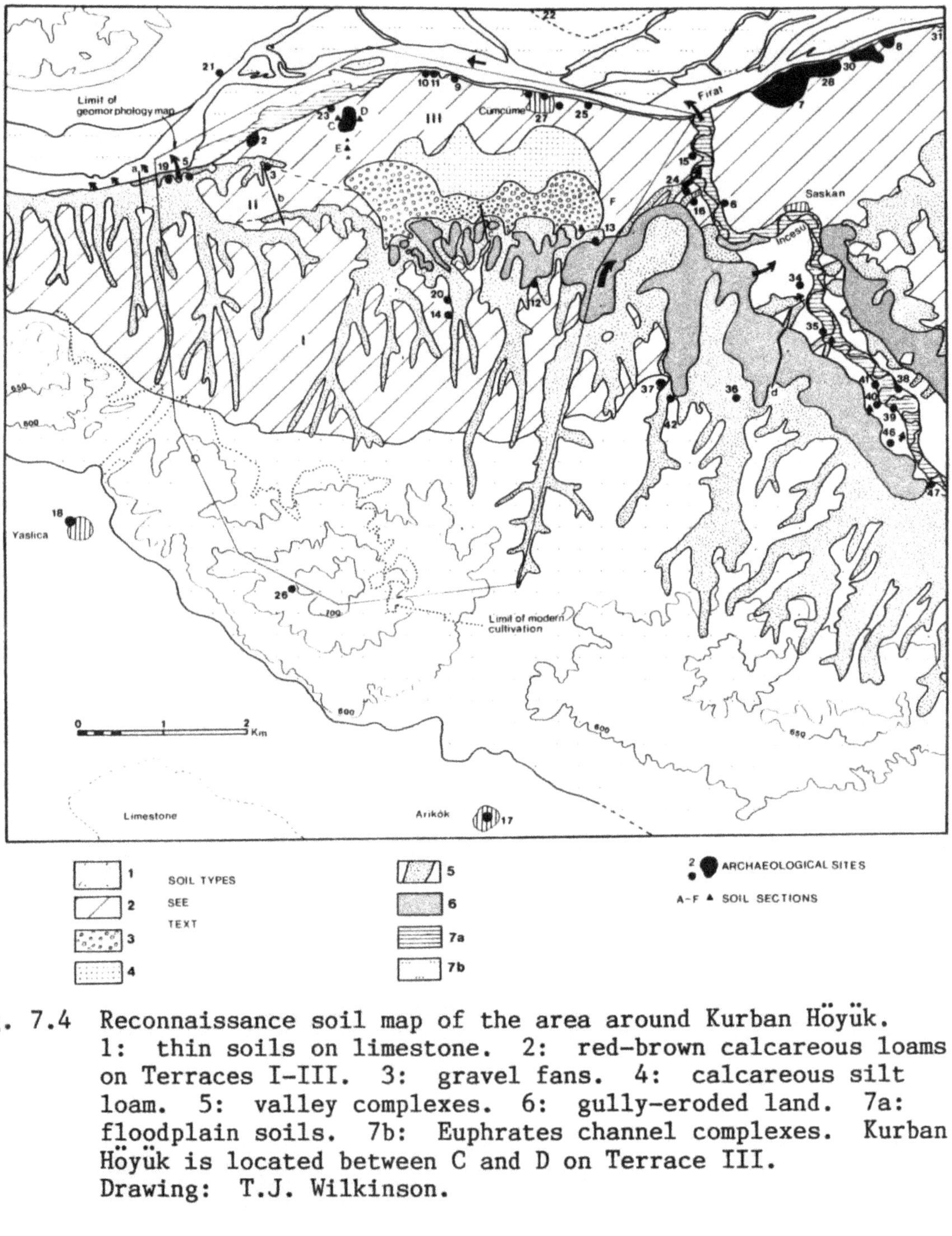

Fig. 7.4 Reconnaissance soil map of the area around Kurban Höyük. 1: thin soils on limestone. 2: red-brown calcareous loams on Terraces I-III. 3: gravel fans. 4: calcareous silt loam. 5: valley complexes. 6: gully-eroded land. 7a: floodplain soils. 7b: Euphrates channel complexes. Kurban Höyük is located between C and D on Terrace III. Drawing: T.J. Wilkinson.

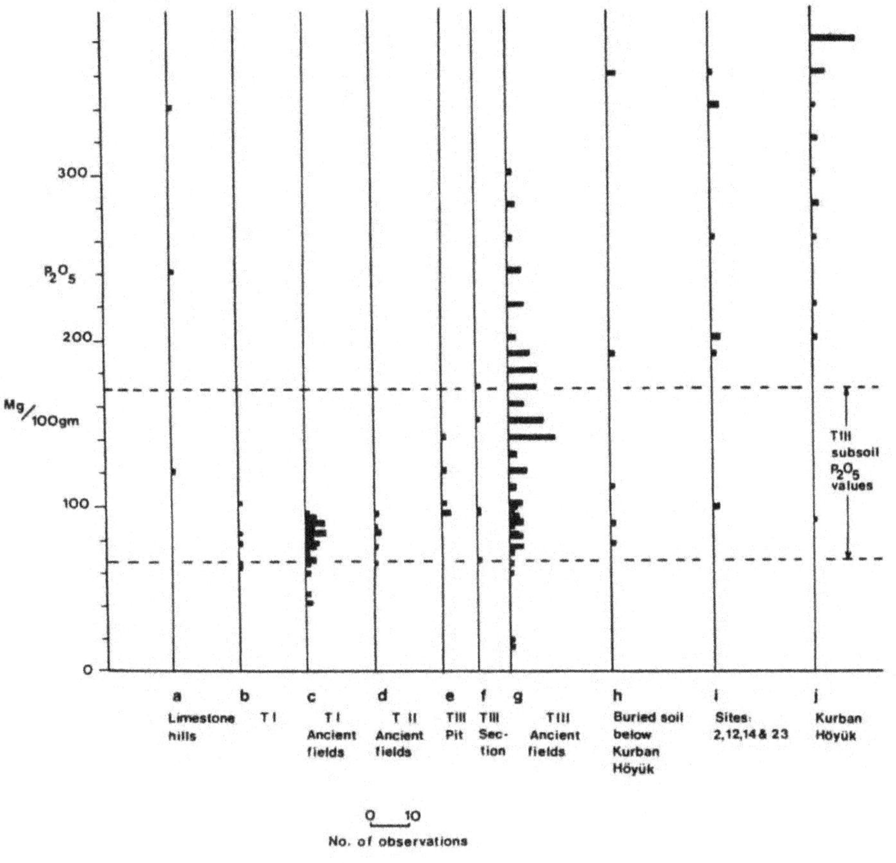

Fig. 7.5 Soil phosphate values plotted according to context. TI-III = Terraces I-III as shown on fig. 4.

f) Five samples from a Type 2 soil profile in Terrace III sediments exposed in bluffs overlooking the Euphrates flood plain (G, fig. 7.6b).

g) 105 samples from artefact-enriched ploughsoils on Terrace III. Mainly Type 2 soils, but also included calcareous silt loams washed from the dissected valleys to the south, (Soil Type 4).

h) Reddish-brown Type 2 palaeosol developed on Terrace III sediments sealed beneath 7m of occupation deposits below Kurban Höyük.

i) Ploughsoils developed in occupation deposits on three late Roman - early Byzantine and one medieval site (Sites: 2, 12, 14 and 23, fig. 7.4).

j) Various occupation deposits from Kurban Höyük.

By sampling a wide variety of contexts, a range of probably 'unenriched' soils (a, b and part of profiles e, f and h) could be compared with soils enriched in phosphates, supposedly either by manuring (c, d and g) or by in situ occupation (part of profile h, also i and j). Although the resultant picture was far from clear, it can be summarised as follows (fig. 7.5):

Shallow soils on limestone (a) exhibited a wide range of values, the highest of which might be accounted for by the presence of phosphate-rich minerals, such as apatite, within the bedrock. Ploughsoils near Sites 12 and 14 on Terrace I (c) although slightly enriched in artefacts (see fig. 7.7), showed negligible phosphate enrichment, as did those of Terrace II (d). Ploughsoils on Terrace III showed a wide range of phosphate values which were commonly in the 150-200 mgP/100gm (mg phosphorus per 100gm of soil) on the Type 2, reddish-brown loams. Although 'control' samples from Type 2 buried soils were mainly in the range 68-120mgP/100gm for apparently un-enriched soils, a peak value of 170mgP/100gm from the lower profile (f) suggests an overall range of 68-170mgP/100gm for the apparently un-enriched soils (fig. 7.5: dashed lines). Anomalously low values on fig. 7.6a (less than 50mgP/100gm) SE and SW of Cümcüme village, occurred on recently deposited Type 4 sediments eroded from Cretaceous limestone slopes immediately to the south.

When viewed in plan (fig. 7.6a) high soil phosphate values occur on Type 2 soils which are evidently more extensive over the western parts of Terrace III. Higher values occur to the north of the terrace and in the vicinity of Sites 2, Kurban Höyük and Cümcüme village (Site 27). As expected, maximum phosphate values, mainly greater than 200mgP/100gm were from occupation deposits at archaeological sites (i and j).

7.4.1 Interpretation of the phosphate results

With the exception of those soils from occupation deposits, only soils of group g, from within the artefact scatters on Terrace III,

a

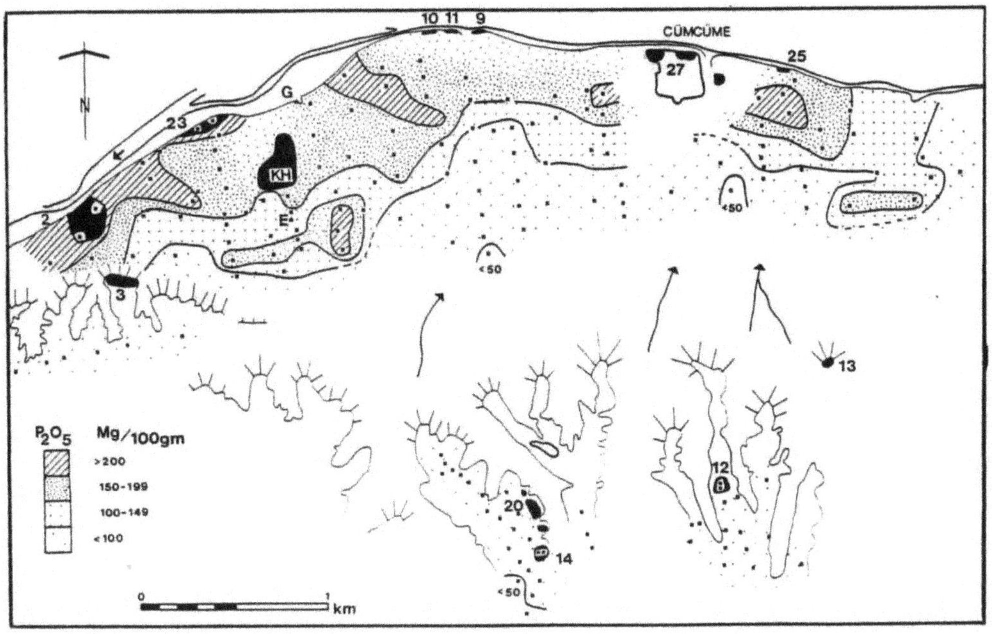

Fig. 7.6 a) Soil phosphate values plotted for Terraces I-III. Sites are numbered and in black; KH = Kurban Höyük.

b

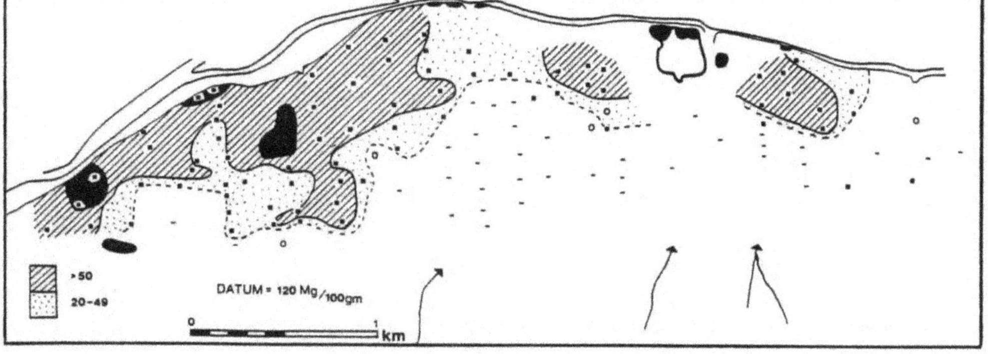

b) Phosphate values exceeding those of the subsoils and buried soils. Drawing: T.J. Wilkinson.

showed phosphate values significantly higher than the control samples. Nevertheless, variations in soil parent material must account for some of the variations illustrated on fig. 7.5. Hence younger calcareous silt loams (Type 4) form lobes of low phosphate values, whereas older, apparently stable soils of Type 2 show higher values. In turn, the stability of the latter soils implies that they were available for the accumulation of phosphates for a longer period than more recent sediments.

In order to isolate phosphate-enriched soils, an arbitrary 'pre-agricultural' value of 120mgP/100gm has been subtracted from the recorded values shown on fig. 7.6a. Positive residuals of 20-49.9 and greater than 50mgP/100gm above this figure were then plotted (fig. 7.6b). Because the maximum subsoil phosphate value of 170mgP/100gm corresponds to a residual value of +50, only those values above this figure can be viewed as being 'probably enriched'. Such soils do, however, appear to cluster around the three sites on Terrace III: Site 2, Kurban Höyük and Site 27.

Although in part still reflecting variations in soil parent material, these areas of peak values may result from sustained use of land by both cultivators and pastoralists. The high soil phosphate values around Kurban Höyük do not however correspond to high sherd densities, which instead focussed upon the late Roman-early Byzantine sites 2 and 27. The latter settlements did exhibit their own surround of higher phosphate values which overlapped roughly with the areas of higher sherd densities. These were of predominantly late Roman-early Byzantine type. A minor component of sherds were of later third millennium type and these were concentrated on a tract of some 80 ha of land to the east of Kurban Höyük (Wilkinson forthcoming). Therefore, the combined phosphate and sherd scatter data can be tentatively interpreted as follows: higher phosphate values corresponding to higher sherd densities around the two late Roman-early Byzantine sites may be the result of manuring with settlement-derived organic material. Higher phosphate values in areas of lower sherd density scatter near Kurban Höyük may result from the addition of phosphate-rich but artefact-poor materials to the soil. Such sources might include stable manure or even the long-continued pasturing of animals around the settlement at night or during certain times of the year. Less clear is why the sparse, but still significant sherd scatters around Sites 12 and 14 did not correspond to higher phosphate values (fig. 7.7). This might be accounted for by the brief duration of manuring around these sites, which, according to surface pottery, were only significantly occupied during the 6th century AD.

It is evident that, although it is possible to make some sense of the soil phosphate data, fluctuations in soil parent material as well as in the source of manure, have resulted in a highly complex pattern of values. These appear to reflect long-continued processes of accumulation of organic waste products. In contrast, artefact scatters proved to be very specific chronologically, being limited to two phases of a 14 phase site chronological sequence extending over 8000 years. Of these two phases, that illustrated on fig. 7.7, was by far the most significant both in terms of artefact density and aggregate site area.

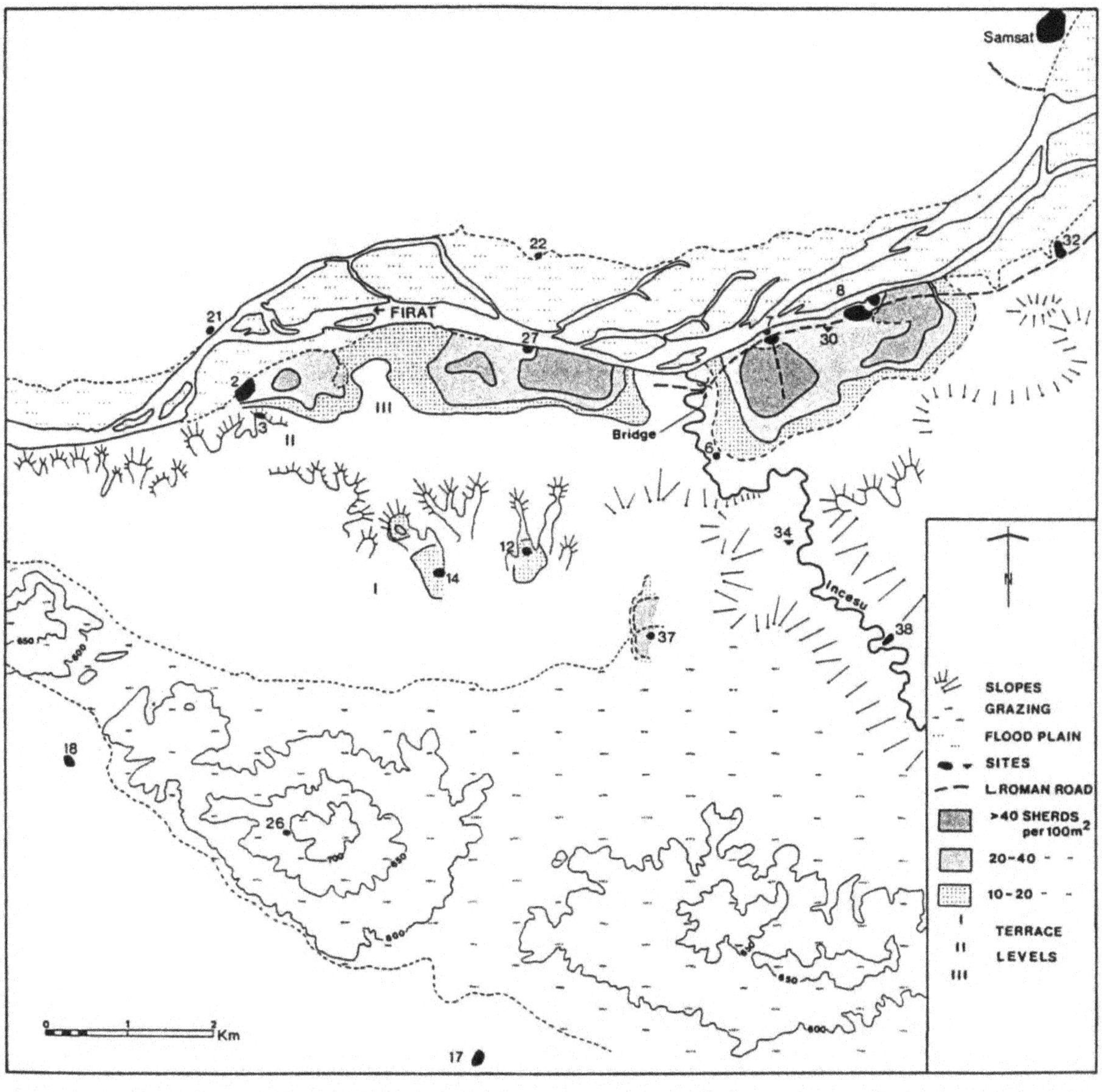

Fig. 7.7 Artefact scatters and other landscape features belonging to the Late Roman-Early Byzantine phase of settlement in the Kurban Höyük area. Drawing T.J. Wilkinson.

7.5 Discussion and conclusions

The results described in this paper show that, irrespective of the reason for their occurrence, artefacts can form significant spreads over extensive tracts of land. They can thus be used as a datum from which episodes of aggradation, erosion and pedogenesis can be measured.

From the perspective of land use, 'field scatters' can most readily be explained as resulting from the use of settlement-derived manure or compost on fields located within an economic haul of settlements. In some cases, but not those examples quoted here, it is possible that pre-existing sites have been dug into to provide fertile soil to spread on fields (Whitcomb, forthcoming). Also there remains the problem of distinguishing 'field scatters' from true habitation sites. In the case of the detailed survey area around Kurban Höyük, this did not present a problem, as habitation sites were presented either by topographic mounding or, more importantly, by spreads of broken limestone foundation stones as well as other indicators (Wilkinson, forthcoming).

Soil phosphates showed some evidence for the prolonged enrichment of soils with phosphate-rich organic materials, but they also provided ambiguous results which reflect both the history of soil development and past land use practice.

In contrast, artefact scatters were shown to have been produced during relatively brief episodes of time. At Siraf and Sohar they were dominantly of early Islamic date; at Tell Sweyhat late 3rd millennium pottery forms dominated the sampled assemblage, and at Kurban Höyük, although a minor early Bronze Age/middle Bronze Age scatter was noted, the main areas of scatter belonged to the late Roman-early Byzantine period. In all the above cases, artefact scatters appeared during a regional peak in population. This was conspicuously the case at Siraf and Sohar where the cities grew in response to the development of maritime trade between Iraq, the Gulf, East Africa, India and China during the heyday of the Abbasid Caliphate (Whitehouse 1968, Williamson 1973). A similar relationship between population and land use could be demonstrated in SE Turkey where detailed survey techniques enabled a demographic curve to be produced for the last 8000 years (Wilkinson, forthcoming). The minor early/middle Bronze Age field scatter corresponded approximately in time to a peak in population during the second half of the third millennium BC. More convincingly, the main scatters corresponded to a substantial peak in population which ensued between the 4th and 6th centuries AD. A similar case can be made for the area around Tell Sweyhat where preliminary survey evidence has shown that population probably reached its peak during the later third millennium BC at which time, a network of tells, in some cases linked by 'hollow way' routes, developed.

Probably linked to the above demographic explanation is one that views the use of settlement-derived manure as part of a strategy employed in the face of a decrease in the availability of animal manure. At Kurban Höyük, a range of indirect palaeoenvironmental and palaeoeconomic indicators, including excavated faunal remains, carbonised seeds and wood, and regional geomorphology, suggested that Holocene woodlands were progressively depleted during the 4th and 3rd millennium BC. By the mid 1st millennium AD virtually all remaining woodland must have gone, and the area was primarily cultivated, except for areas of rough pasture which remained on the limestone uplands and the Euphrates flood plain (fig. 7.7 and soil types 1 and 7, fig. 7.4). During the earlier periods characterised by greater amounts of available woodland, firewood was readily available as domestic fuel. As this source diminished, greater reliance was placed upon the use of animal dung as fuel; this remains the practice today, (Miller and Smart 1984). Consequently, during times of population growth, when land use intensification was necessary, there was also greater competition for manure and some manures or composts had to be procured from alternative sources: cess pits, street sweepings etc.

The appearance of extensive manuring-related sherd scatters at times of population peaks lends some support to Boserup's hypothesis that, increasing population density is the major cause of increased agricultural intensity (Boserup 1965, see also Grigg 1982, 37 and Table 7.1). Within this context it is significant that all the above population peaks appeared to result from exogenous causes.

In the case of the Kurban Höyük area, population increase was probably linked with vegetation decline which, in turn affected the availability of manuring resources. Furthermore, the depletion of vegetation cover also initiated major episodes of accelerated erosion and aggradation. This resulted in the loss of topsoil from large areas and its accumulation as alluvial fans on nearby river terraces.

This study has concentrated upon the positive evidence of field scatters. As a counterweight, it must be emphasised that such scatters do not occur universally. Instead, they seem to result when a specific range of conditions prevailed. Nevertheless, although still in a preliminary stage of study, it should be evident that field scatters provide a tantalizing and highly instructive link between demographic and ecological history and pedogenic development.

Acknowledgements

I wish to thank Paolo Costa (Department of Antiquities, Muscat, Sultanate of Oman) and His Highness Sayyid Faisal bin Ali bin Faisal, (Minister of National Heritage, Sultanate of Oman); David Whitehouse (Corning Glass Museum, New York); T.A. Holland (Oriental Institute, University of Chicago), and L. Marfoe (University of Chicago), for their permission to publish data in advance of final publication. I am especially grateful to David Gurney of the Norfolk Archaeological Unit, for providing the necessary data for phosphate analysis. I also wish to thank M.L. Ingraham (University of Toronto), D. Whitcomb (Oriental Institute, Chicago) and N. Ralph (University of Sheffield) for providing information or comments which contributed to this article.

References

Boserup, E., 1965. The conditions of agricultural growth. London: George, Allen and Unwin.

Chisholm, M., 1979. Rural settlement and land use. London: Hutchinson.

Costa, P.M., 1978. The copper mining settlement of 'Arja: a preliminary survey. J. of Oman Studies 4, 9-14.

Costa, P.M. and Wilkinson, T.J., forthcoming. The hinterland of Sohar. J. of Oman Studies, monograph.

English, P.W., 1966. City and village in Iran. Madison, Wisconsin.

F.A.O.-Unesco, 1977. Soil map of the world 7. South Asia. Paris.

Fowler, P.J., 1981. Later prehistory. In: S. Piggot, (ed.), The agrarian history of England and Wales, 1.1. Cambridge, 61-298.

Gaffney, U. and Tingle, M., 1985. The Maddle Farm (Berks) project and micro-regional analysis. In: S. Macready and F.H. Thompson (eds.), Archaeological field survey in Britain and abroad. Society of Antiquaries, occasional paper (new series) 6, London, 67-73.

Gorbunov, B.V., 1965 (1968). Irrigated soils of Central Asia. In: V.A. Kovda and E.V. Lobova, (eds.), Geography and classification of soils of Asia, 34-45. Israel Programme for Scientific Translations, Jerusalem.

Greene, J.A., 1984. Canadian Carthage survey, 1983. Echos du monde classique, classical views 28 (new series 3), 214-218.

Grigg, D., 1982. The dynamics of agricultural change. London: Hutchinson.

Hamond, F.W., 1983. Phosphate analysis of archaeological sediments. In: T. Reeves-Smith and F.W. Hamond, (eds.), Landscape archaeology in Ireland. B.A.R., Brit. Ser. 116, 47-80.

Holland, T.A., 1976. Preliminary report on excavations at Tell es-Sweyhat, Syria, 1973-74. Levant 8, 36-70.

Hughes, M.J., 1976. Atomic absorption techniques in archaeology. Archaeometry 18, 19-37.

ILACO., 1975. Water resources development project, northern Oman. Ministry of Communications, Water Resources Centre, Muscat, Oman.

Lattimore, O., 1979. Geography and the ancient empires. In: M.T. Larsen (ed.), Power and propaganda, a symposium on ancient empires. Copenhagen: Akademisk Forlag, 35-40.

Marfoe, L., 1986. The Chicago Euphrates archaeological project 1980-84: an interim report. Anatolica. Forthcoming.

Miller, N. and Smart, T.L., 1984. Intentional burning of dung as fuel: a mechanism for the incorporation of charred seeds in the archaeological record. J. of Ethnobiology, 4, 15-28.

Murphy, J. and Riley, J.P., 1962. A modified single solution method for the determination of phosphate in natural waters. Analytica Chimica Acta 27, 31-36.

Rhodes, P., 1950. The Celtic field systems on the Berkshire downs. Oxoniensia 15, 1-28.

Semple, E.C., 1932. The geography of the Mediterranean region. London: Constable.

Sieveking, G. de G., Longworth, I.H., Hughes, M.J., Clark, A.J. and Millet, A., 1973. A new survey of Grimes Graves Norfolk. Proceedings of the Prehistoric Society 39, 182-218.

Subramanian, V., Zarif-Salem, M. and Nasyrov, Y., 1965 (1968). Data on the soils of Afghanistan. In: V.A. Kovda and E.V. Lobova (eds.), Geography and classification of soils of Asia. Israel Programme for Scientific Translations, Jerusalem.

Whitcomb, D.S., in press. Islamic archaeology in Aden and the Hadhramaut. In: Festschrift for Beatrice de Cardi.

Whitehouse, D., 1968. Excavations at Siraf: first interim report. Iran 6, 1-22.

Wilkinson, T.J., 1974. Agricultural decline in the Siraf region, Iran. Paleorient 2.1, 123-132.

Wilkinson, T.J., 1975. Sohar Ancient Fields Project, interim report no. 1. J. of Oman Studies 1, 159-166.

Wilkinson, T.J., 1976. Sohar Ancient Fields Project, interim report no. 2. J. of Oman Studies 2, 75-80.

Wilkinson, T.J., 1977. Sohar Ancient Fields Project, interim report no. 3. J. of Oman Studies 3, 13-16.

Wilkinson, T.J., 1982. The definition of ancient manured zones by means of extensive sherd-sampling techniques. J. of Field Archaeology 9, 323-333.

Wilkinson, T.J., forthcoming. Settlement and land use in part of SE Turkey. In: L. Marfoe (ed.), Town and country in early south-eastern Anatolia 2. Chicago: Oriental Institute.

Williams, D., 1973. Flotation at Siraf. Antiquity 47, 288-292.

Williamson, A., 1973. Harvard Archaeological Survey in Oman 1973: Sohar and the sea trade of Oman in the 10th century AD. Proceedings of the Seminar for Arabian Studies 4, 78-96.

Williamson, T.M., 1984. The Roman countryside: settlement and agriculture in NW Essex. Britannia 15, 225-230.

8. ASPECTS OF BURIED PREHISTORIC SOILS IN THE LOWER WELLAND VALLEY AND THE FEN MARGIN NORTH OF PETERBOROUGH, CAMBRIDGESHIRE

C.A.I. French

Abstract

Ten sites/landscapes are examined pedologically in the lower Welland valley, and the fen margin and fen basin of Cambridgeshire. In the lower Welland valley, remnants of pre-Neolithic buried soils occur as argillic brown earths, which probably developed under stable woodland conditions. The present day soils are brown earths, which are often decalcified. The buried prehistoric soils of the floodplain and fen margin are brown earths of varying degrees of maturity, which are now overlain by alluvium and/or peat. The buried pre-Neolithic soils of the fenland basin are argillic brown earths which developed on dry land under forested conditions prior to the advent of the freshwater and marine flooding episodes which gave rise to the East Anglian fens.

8.1 Introduction

The study area is situated in north-west Cambridgeshire at the confluence of the lower Welland and lower Nene valleys and the East Anglian fenland basin (figs. 8.1 and 8.2). Ten sites/landscapes within the study area are examined in this paper. The sites of Maxey (3) and Barnack/Bainton (9) are located in the lower Welland valley on first terrace sand/gravel drift deposits. Although both sites at Etton (1 and 2) are situated on a similar subsoil they are fen margin sites buried beneath blanketing alluvial deposits. By way of comparison the sites at Haddenham (10) are on fen margin river terrace deposits in the Great Ouse river valley in southern Cambridgeshire, buried beneath peat and alluvial deposits. Within the fenland itself, the sites of Crowtree Farm (4), Oakhurst Farm (5), Guy's Fen (6), Borough/Newborough Fens (7) and Flag Fen (8) are discussed in relation to their appropriate place in the fenland sedimentary sequence.

Buried soils have been revealed through a combination of rescue excavations and planned dyke survey undertaken by the Fenland Archaeological Trust under the direction of Dr. F.M.M. Pryor and the author since 1979. Throughout, soil micromorphological techniques have been used (after Bullock et al. 1981) in conjunction with particle size analysis (after Shackley 1975) and molluscan analysis (after Evans 1972). As most of the work in the fenland region has just begun, the results to date must be considered as preliminary in nature.

The sites will be discussed in approximate chronological order, first considering the lower Welland valley and the fen margin sites,

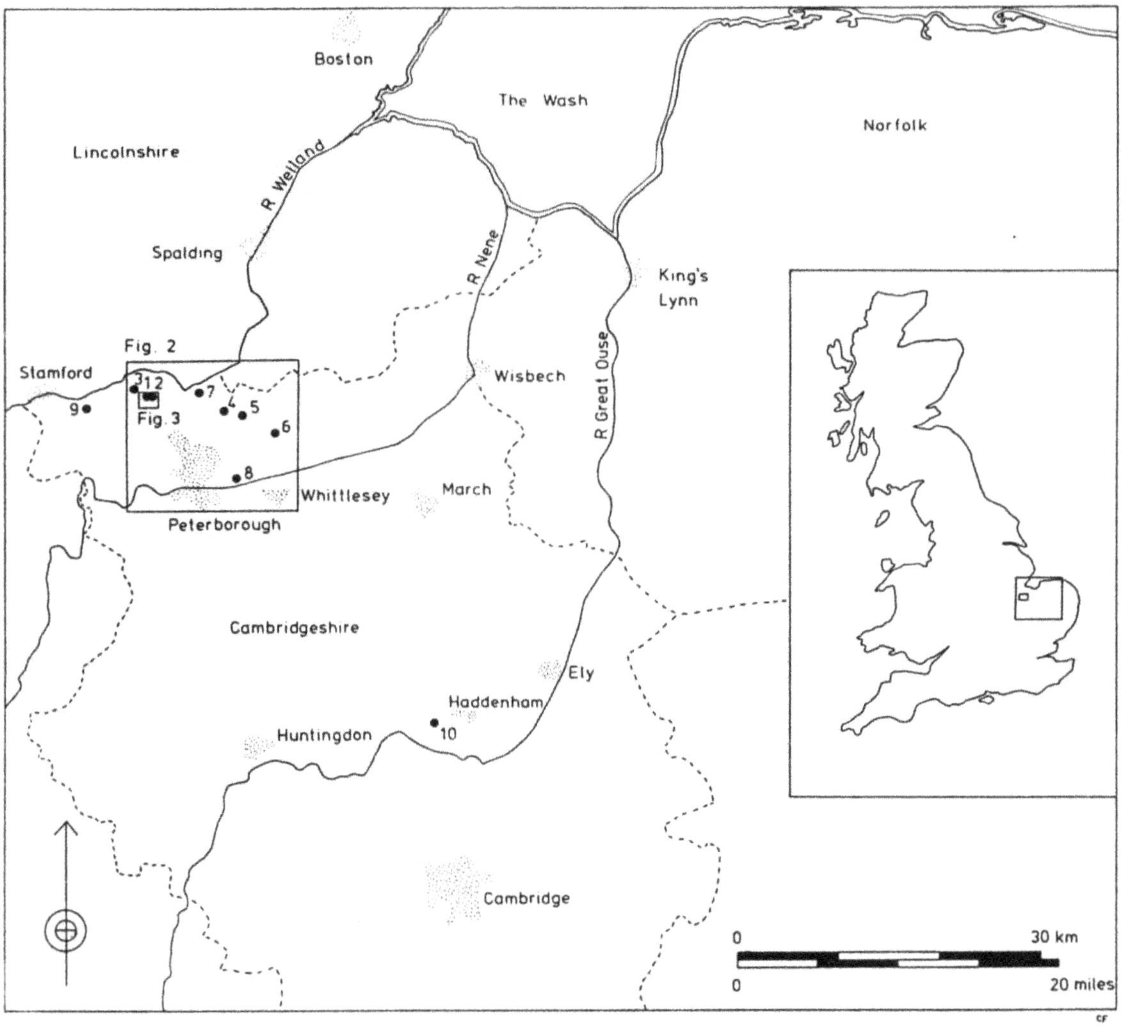

Fig. 8.1 General map of East Anglia illustrating the location of the sites discussed in the text. 1: Etton Woodgate; 2: Etton; 3: Maxey; 4: Crowtree Farm; 5: Oakhurst Farm; 6: Guy's Fen; 7: St. Pega's, Borough Fen; 8: Flag Fen; 9: Barnack/ Bainton; 10: Haddenham (Drawing: C.A.I. French).

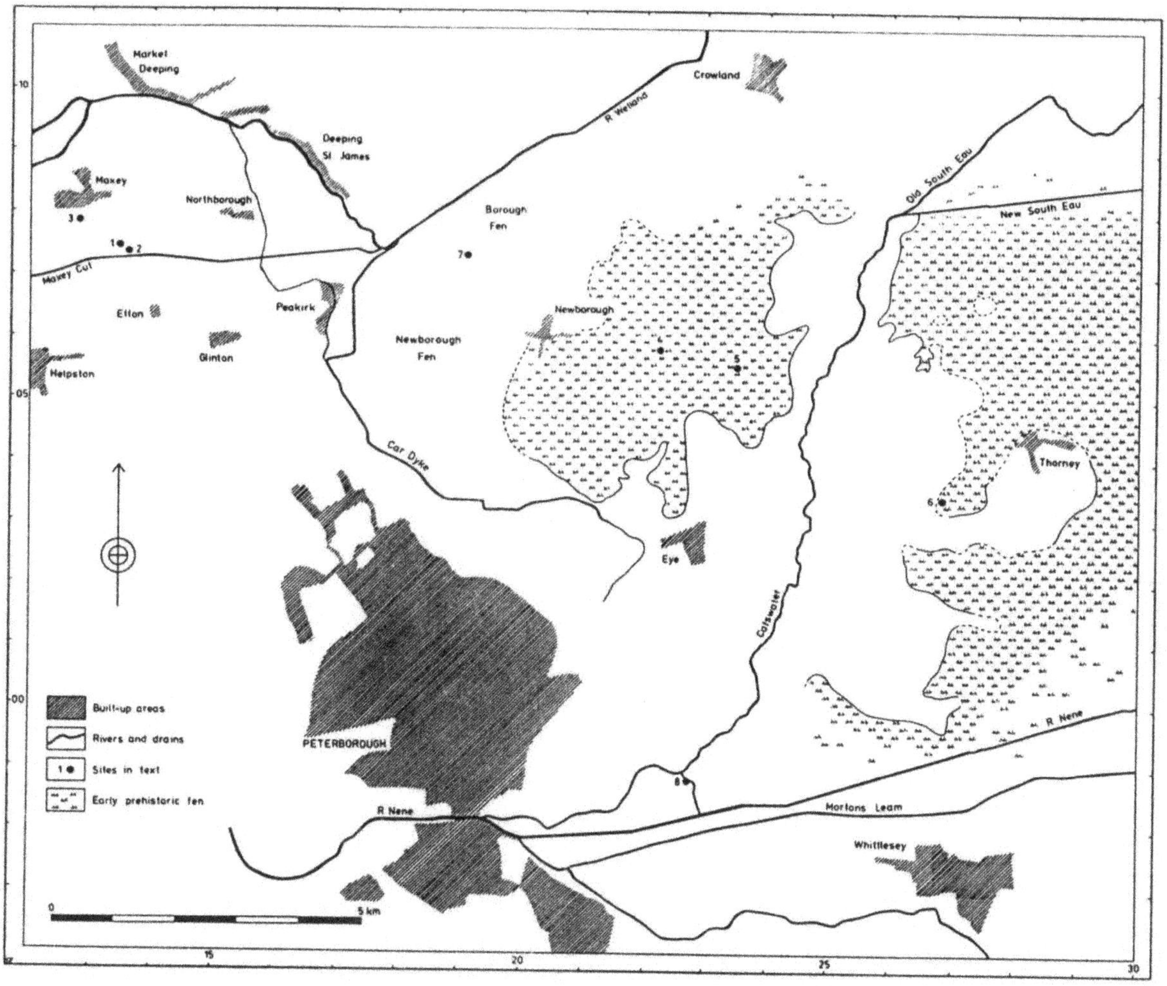

Fig. 8.2 Map of the lower Welland valley and the fen margin in the immediate vicinity of Peterborough. The approximate extent of the late Neolithic fen margin is indicated. Sites 1 to 8 are shown (as for fig. 1) (Drawing: C.A.I. French).

and then the preliminary investigations in the north-western and southern fenland of Cambridgeshire.

8.2 The Lower Welland Valley and Fen Margin

8.2.1 Etton Woodgate (1)

Quarrying operations in the Maxey gravel pit (Tarmac Ltd.) revealed an enormous early Neolithic ditch defining the eastern edge of the higher ground of Maxey "island" (fig. 8.3). A probable rectilinear post-built structure of the same period was situated within its single entranceway. Also, the manufacture of oak charcoal possibly took place here during the earlier Bronze Age (Pryor et al. 1985). The site was obscured beneath ca. 50-80cm of silty clay alluvium, thought to have been deposited since late Iron Age times (French 1983).

The south-eastern corner of the site is bounded and partially eroded by a series of former meandering stream channels. The lower fill of the stream channels is composed of a locally derived accumulation of silty clay and sand from the subsoil. The upper fill on the other hand exhibits a dense loamy fabric, with coarser and finer zones. This suggests that it was not homogenized after burial. There is also a high silt and organic matter content which may be partially due to seasonal stream contribution. Together these characteristics suggest that this fabric is colluvium, locally derived from the higher ground of Maxey "island" immediately upstream and to the west. Derived material also infills some of the pore space, and is suggestive of the continued addition of eroded material. The whole matrix has also been subject to alternating wet and dry periods, and is indicative of seasonal waterlogging.

The buried soil preserved beneath the alluvium is an argillic brown earth with a gleyed lower B horizon. The A and upper B horizon fabric are of a mixed nature: the more homogeneous, clayier areas of the fabric are indicative of an alluvial component to the soil, whereas the less homogeneous areas of mixed sand, silt and clay are suggestive of a colluvial component. The very abundant charcoal content and the mixing of material into the soil probably results from the manufacture of charcoal and human occupation, respectively, in the immediate vicinity. The fabric is ferruginised, which is indicative of wetting and drying, and seasonal waterlogging. As this soil does not appear to have suffered any major truncation by man or by the later phase of alluviation, the most reasonable explanation is that the colluvium found in the upper fill of the stream course covered a much greater area of the prehistoric soil, and which has since been eroded and incorporated with the overlying later alluvium.

8.2.2 Etton (2)

The middle Neolithic causewayed enclosure at Etton is located ca. 75m to the east of Etton Woodgate (fig. 8.3). It encloses about eight acres (ca. 3.4 ha), more or less defining a low gravel knoll, which is bounded by a series of stream channels on its north-western side. These channels were active both before, during and after the use of the

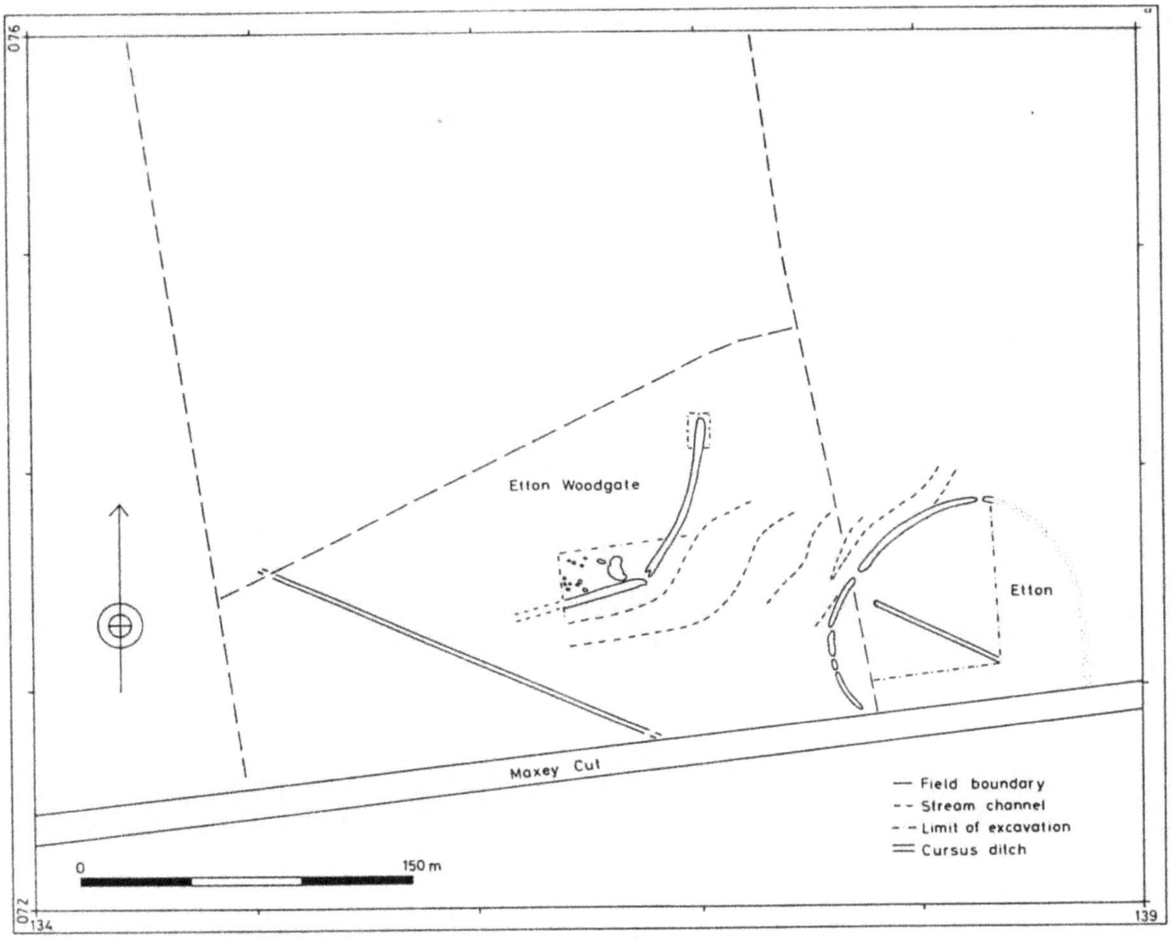

Fig. 8.3 A sketch plan of the sites of Etton Woodgate and Etton, showing the intervening stream courses (Drawing: C.A.I. French).

enclosure ditch.

This site is also buried beneath ca. 50-130cm of silty clay alluvium which has probably been deposited from late Iron Age times until 1953 when seasonal flooding ceased following the enlargement of the nearby Maxey Cut. It is estimated that there may have been a minimum of ca. 0.04 - 0.05cm per year of accumulation over this lengthy period (French 1983).

Sealed beneath the alluvium is the pre-late Iron Age soil. It is a partially developed brown earth subject to gleying. Micromorphology suggests that this soil is a slightly truncated organic lower A horizon resting on a weathered B/C horizon. Although this soil shows no signs of anthropogenic disturbance, the upper part of the A horizon has probably been truncated prior to the deposition of the alluvium. Indeed the high silt content suggests a possible alluvial contribution to the upper part of the profile. The replacement of organic matter by oxidised iron indicates that this soil has been subject to seasonally wet and dry conditions, both before, during and after the use of the enclosure (French in Pryor et al. 1985).

The preliminary pollen analysis of the primary peat deposits in the enclosure ditch suggests that the area was significantly open by the middle Neolithic period, and that aquatic and semi-aquatic plants predominated (Scaife 1983). Although cereal pollen was found in the ditch (Scaife 1984), there is no definite evidence of cultivation. Nevertheless, excavation of the interior of the enclosure in 1985-7 has revealed the division of the enclosure into two halves.

The secondary deposits in the enclosure ditch consist of a mixture of finer material (sandy clay loam) and coarser material (sandy loam, fine and small gravel pebbles). This combination of infilling material is suggestive of eroded material becoming entrapped in the ditch. The lower secondary deposits contain large quantities of wood, including coppiced rods and wood-working debris (M. Taylor, pers. comm.). The wood has been preserved because of waterlogging due to a high ground water table as well as the partial exclusion of oxygen by the overlying cover of alluvium. But the site has been de-watered since the summer of 1983 due to quarrying operations. This has caused a rapid and immediate deterioration in the preservation of the wood (noticeable within one month) and the other macro-botanical remains contained within the ditch (French and Taylor 1985).

The tertiary fill of the ditch consists of a slow natural accumulation of sandy to sandy clay loam. The depression caused by the ditch probably held freshwater seasonally, possibly giving an alluvial contribution to the uppermost ditch fill.

8.2.3 Maxey (3)

About 0.5km to the west on the higher ground of Maxey "island" lies an enormous complex of prehistoric and Roman sites (RCHM 1960; and see fig. 40 in Pryor and French 1985). The focus of Maxey comprises a later Neolithic henge containing a central ring ditch with a two-phase mound, and a small oval barrow situated in its entranceway (Pryor and

French 1985). Fortunately, a mediaeval plough headland situated over both mounds insured the preservation of both mound material and the underlying buried soil. Otherwise on this higher ground the soil is unprotected and subject to intense arable cultivation. Micromorphological analysis of the protected remnants of the buried soils and the make-up of the mounds enabled both the interpretation of the archaeological complex as well as the elucidation of the soil history associated with these monuments.

The small oval barrow was built in two stages. The first consisted of a continuous oval of vertical, square-dressed oak timbers placed edge to edge, probably acting as a mortuary enclosure. The second consisted of a low mound with a central crouched inhumation built over the now burnt down mortuary enclosure. This mound consisted of redeposited A and B horizon material. No turves were evident (French 1983; Pryor and French 1985).

The underlying buried soil was a truncated, argillic brown earth with an upper, thicker, coarser textured zone and a lower, thinner, finer gleying textured zone. Both zones exhibited a Bt fabric, as well as signs of post-burial gleying. Subsequently, this soil suffered a deep truncation, probably associated with and immediately prior to the construction of the mortuary structure. The abundance of cutans with strong continuous orientation indicates the illuviation of clay under former stable, probably wooded conditions. There were also a few possible agricutans containing fine charcoal and organic matter which are suggestive of limited soil disturbance, possibly due to agriculture.

Heavy mineral analysis of this buried soil revealed a relatively high proportion of collaphane, the dominant mineral of fossil bone in which it is formed by phosphatic enrichment (French 1983). So, although no bones were found within the mortuary structure itself, it may just possibly have been used as an ossuary (Pryor and French 1985).

The central mound within the henge was also built in two stages. Its first phase comprised a small turf-built mound, berm, internal gravel bank and continuous ring ditch. The bank was later partially slighted into the ditch, and a large mound was constructed of redeposited A and B horizon material dumped over and incorporating the bank and primary turf mound.

The buried soil was only preserved beneath the primary mound. It had suffered some truncation, but leaving the lower A horizon undisturbed. This lower A horizon exhibited three zones or laminations as well as a high silt content, which are suggestive of a considerable loessic component (French 1983). It may result from erosion processes during the late Devensian and early Flandrian periods, but the relative clarity of the laminations is suggestive of recent wind-blow caused by early agriculture (Weir et al. 1971; Catt 1978). The presence of dirty argillans is suggestive of clearance (Macphail, in press) and possibly represents some early phases of cultivation (Bouma 1969; Kwaad and Mücher 1979), but not of a severe or prolonged nature.

In summary, the henge complex (henge, central and oval mounds) underwent four major phases of development. First, prior to the late

Neolithic period an argillic brown earth had developed, became deforested and underwent some agricultural disturbance. Corroborative evidence is provided by the building of lengths of cursus ditches (c. 4km in total length; both cursus ditches are cut by the henge ditch) earlier in the later Neolithic period, an activity which must be viewed in the context of an open landscape. The pollen evidence from the nearby middle Neolithic causewayed enclosure at Etton also provides complementary evidence to this argument. Second, the henge, central primary mound and mortuary structure were designed and built as a unitary whole (Pryor and French 1985). The soil suffered a deep truncation prior to the building of the last two structures. Third, the henge ditch was partially back-filled, the mortuary structure was burnt down, and the bank surrounding the central mound was slighted. Finally, the oval mound and secondary central mound were constructed.

8.2.4 Barnack/Bainton (9)

Further evidence for the opening-up of the landscape in the lower Welland valley was found ca. 4km upstream from the Maxey sites between the modern villages of Bainton and Barnack (fig. 8.1). The present soil is a decalcified brown earth overlain by alluvium in places.

Here a buried soil of argillic brown earth type was preserved beneath a two-phase mediaeval plough headland, and associated with late Neolithic/Beaker period activity. The soil exhibited a lower A/upper B fabric with some characteristics of a Bt horizon. There were two fabrics - silt/clay aggregates with abundant charcoal and without charcoal. The former fabric, plus the presence of many dusty agricutans, suggest that the soil was physically mixed by cultivation prior to the late Neolithic activity. The latter fabric suggests that this was followed by a stable phase. Certainly by the later Iron Age, molluscan evidence from adjacent drainage ditches suggests open meadow-land in the immediate vicinity (French 1983). Alluvium had begun to accumulate by the Roman period, as at Etton, and probably would have made the area unusable except for seasonal pasture (French 1983; Pryor and French 1985).

Thus by the end of the Neolithic period, there is every indication that the valley floor of the Welland was a largely cleared landscape. Colluvium, presumably mainly as a result of clearance and agriculture, had begun to accumulate downstream in low-lying areas. By the beginning of the 1st millennium AD, alluvium had begun to be deposited in most low-lying areas, especially around Maxey "island" and eastwards towards the fen margin. This must mainly be due to a further intensification of clearance and agriculture on the higher ground to the west and on the valley slopes.

8.3 <u>Haddenham</u> (10)

Many sites in the parish of Haddenham in southern Cambridgeshire in the lower Great Ouse river valley (fig. 8.1) have been discovered by D.N. Hall for the Fenland Project, and several sites have been excavated by C. Evans and Dr. I. Hodder (1983/4; 1984/5). These sites include a causewayed enclosure, barrow and Romano-British enclosure

(Evans and Hodder 1983/4, fig. 8; 1984/5, fig. 6).

The pre-peat buried soil at the causewayed enclosure site was a slightly truncated, gleyed, weakly developed argillic brown earth. It contained numerous rolled clay aggregates indicative of alluviation at some stage. The pre-barrow soil on the other hand was a very immature, gleyed brown earth which had undergone little soil development, but was apparently free of alluvial influence.

The buried soil beneath the Romano-British enclosure bank comprised two horizons. The upper horizon, an oxidised sandy loam, contained a mixture of three fabrics: a Bt fabric with clay coatings, a lower A/upper B fabric with amorphous organic matter, and an upper A fabric of sand/peat aggregates. There were also fragments of fine fabric material in the void spaces. All these features suggest that there was some pre-Roman disturbance and possibly cultivation. The lower horizon exhibited a lower A horizon fabric which was rich in organic matter and was somewhat disaggregated. Although this horizon had been disturbed, there was no evidence of cultivation. Both horizons contained abundant rolled clay aggregates indicative of alluviation (French 1984/5, 23-25).

Consequently, alluviation in this part of the Great Ouse valley must have begun at some time after the early Bronze Age, but prior to the early Roman period. During the same period there had also been some peat growth, which had subsequently dried out. These factors, plus little soil development and limited agricultural use, hint at the marginal positioning of these sites on the Great Ouse floodplain.

8.4 The Fenland

Despite the enormous body of work on the botanical and sedimentalogical evidence of the East Anglian fenland basin (Godwin 1975 and refs.), relatively little archaeological work has been carried out until recently. In response to this and the severe deflation and dewatering effects of the draining of the fens since the mid-17th century AD, both extensive and intensive archaeological survey has begun, the former by D.N. Hall for the Fenland Project, and the latter by Dr. F.M.M. Pryor and the author.

Our intensive survey involves following behind the drainage contractors as and when drainage dykes are being cleaned. Lengthy sections through an otherwise invisible landscape are thus observed, recorded, levelled and photographed, and prospecting for archaeological sites and buried soils becomes possible (for rationale see Crowther et al., 1985; Pryor 1985a). Then this new archaeological record may be placed against the appropriate background of the fen succession.

8.4.1 Crowtree and Oakhurst Farms (4 and 5)

So far four late Mesolithic sites have been discovered. Two of the best preserved are Crowtree and Oakhurst Farms situated about 1 and 2km to the east of the village of Newborough, respectively (fig. 8.2). They are situated on small, low "islands" of fen gravel, and are buried by deposits representing the full fen succession: lower peat, fen clay and upper peat.

Although micromorphological analysis of the buried argillic brown earths has yet to be completed, the profiles appear to be more or less undisturbed and intact. When occupied the greater part of these sites were on dry land, and the making of flint tools probably took place on site. In terms of potential preservation, these sites are without parallel in Britain.

8.4.2 Guy's Fen Trackway (6)

On the west side of Thorney "island", a narrow wooden trackway was built between two gravel "spurs", probably in response to the marine incursion responsible for depositing the fen clay in the later Neolithic period. The area of deposition of the fen clay represents an enormous former marine estuary or salt marsh. It is now believed to have been deposited in two major episodes, each with its own system of seaward draining channels or "rodhams". The infilling of the rodhams probably occurred by a major breaching of the existing dune system at the seaward edge of the fenland basin (Dr. J. Zalaciewicz, British Geological Survey, pers. comm.). This event or events may have been more catastrophic than the marine incursion in human terms. Thus throughout this period, human occupation must have been pushed onto the higher ground, the fen "islands" and the fen margins.

8.4.3 Borough and Newborough Fens

Both of these fens would have been on the margin of the former marine estuary/salt marsh represented by the deposition of the fen clay during the later Neolithic, and during the early Bronze Age period (fig. 8.2). This land lies immediately to the northeast of Etton and the city of Peterborough in a ca. 5-6km wide band. Every dyke examined in these two fens has revealed buried, formerly dry land soils, with scattered evidence of archaeological activity. These buried soils are overlain by upper peat and/or alluvial deposits. This buried landscape is preserved over hundreds of acres of land, not just confined to a few profiles.

But preliminary micromorphological evidence suggests that the buried soil is an argillic brown earth which has been severely truncated in places. The two horizons uniformly evident in the field proved to be Ea or upper B horizon and Bt horizon material. Despite severe leaching of the profile, numerous clay coatings, both laminated and non-laminated were evident in the Bt horizon. The numerous clay coatings are indicative of once stable forest conditions pertaining, probably earlier in the Neolithic period. But, coatings of both types appear to have been physically broken-up and have had their orientation changed. This is perhaps a result of soil turbation, possibly caused by freshwater flooding. If this flooding is also the cause of the severe truncation of the soil, we must revise our view of the amount of erosion caused by such events in some circumstances, and its obvious affect on the extant archaeological record.

Fortunately, similar buried soils are preserved beneath many Bronze Age barrow mounds which are located just beyond the western limit of the fen clay. These mounds are now beginning to emerge from the upper peat. Also, similar soils are found on the banks of the Welland first terrace gravels where the river enters the fen basin in the northern part of Borough Fen. Here a mixture of upper peat and alluvium protects the prehistoric soils. At the middle Iron Age fort site of St. Pega's (fig. 8.2, site 7), molluscan evidence from the buried soil is suggestive of damp open conditions (French 1983).

8.4.4 Flag Fen (8)

In other areas of the fen basin, soils have had little chance of formation and development. In Northey Fen, immediately east of Peterborough and just to the north of where the present River Nene enters the fen basin (fig. 8.2), the upper peat probably began to form later in the Bronze Age. In this fen the underlying soil has been under the long term influence of freshwater, thereby preventing much soil development. But in the shallow growing fen a settlement was constructed on a wooden platform (named Flag Fen), of about 1 ha in area, during the late Bronze Age (Pryor 1985b).

8.5 Conclusions

Although this is only a brief summary of the present work in an enormously complicated area, it is evident that there is an extensive variety of buried soils in various stages of development to be examined in the fenland basin. They must be viewed in terms of buried, more or less complete, dry land landscapes at varying stages in the fenland succession.

Despite the paucity of sites with buried soils that have been examined, some general remarks can nevertheless be made about the distribution of prehistoric soils in Cambridgeshire and man's effects on them. In the lower Welland valley where prehistoric soils survive beneath upstanding monuments on flood-free ground on the valley floor, they are argillic brown earths as opposed to the present day brown earths. These former soils were presumably once much more widespread. At the sites of Maxey, Barnack/ Bainton and Etton Woodgate, there is little doubt that pre-late Neolithic clearance heralded the beginning of the change in soil type from argillic brown earth to the brown earths of today. In one instance at Maxey, clearance and cultivation of the soil may have been sufficient to have caused soil movement by wind-blow, as well as the transport and deposition of colluvium in Neolithic period stream channels ca. 0.5km downstream at both the Etton sites.

On the valley floodplain and the fen margin areas, the pre-Bronze Age buried soils are brown earths in various stages of maturity which exhibit varying degrees of gleying, both pre- and post-burial, by later deposits. In part, this may be related to either the absence of forest cover or the lack of stable and prolonged forest cover. Susceptibility of these areas to seasonal flooding may have played as much of a role in hindering soil development as did man's possible activities. Today these soils are generally overlain by alluvium and/or desiccated peat.

In the fenland basin, the pre-peat and pre-marine deposit soils appear to be well-developed argillic brown earths, subject to the effects of later gleying and disturbance by natural erosion processes. Therefore argillic brown earth soils had developed under stable forest conditions over extensive areas of the fenland by the Neolithic period and prior to the formation and deposition of the various fenland deposits. As yet it is too soon to speculate as to man's effects on these soils prior to the drowning of the fens during the Neolithic period. Nevertheless, there may have been some limited clearance and occupation on the small "islands" in the fens as is indicated by preliminary work at the late Mesolithic/early Neolithic sites of Crowtree and Oakhurst Farms.

As this resource is under threat from both drainage and agricultural operations, every effort must be made to preserve and study as much as is possible of these buried "soil-scapes". To this end, the enormity of the task dictates that a "soil bank" must be set up. This will involve the study and preservation for the future of as many fenland soil monoliths as possible.

It is now apparent from the micromorphological study of buried soils on a variety of dry land sites that it is an extremely useful and reliable method of discerning the processes of past and present soil development. These observations often have direct implications for past land use and human activities. But, far too often buried soils on most dry land sites tend only to be preserved in direct association with a prehistoric monument, such as a barrow mound or bank. In many cases this association may bias the interpretation, and more often than not has caused additional soil disturbance, such as severe truncation of the soil profile. But in the fen and fen margin areas, whole buried landscapes are preserved, both in association and not in association with archaeological sites and over differing topographies. Thus, the detailed interpretation of landscapes through time, in response to natural environmental and human factors, is possible. Herein probably lies the most valuable potential archaeological resource in Britain today.

Acknowledgements

I would like to thank Dr. R.I. Macphail and Dr. R.G. Scaife of the Department of Human Environment, Institute of Archaeology, London; Dr. J. Zalaciewicz, British Geological Survey; Dr. F.M.M. Pryor and Maisie Taylor, Fenland Archaeological Trust; and David N. Hall of the Fenland Project.

References

Bouma, J., 1969. Microstructure and stability of two sandy loam soils with different soil management. Agric. Res. Rep. Wageningen, Nr. 724.

Bullock, P., Murphy, C.P. and Waller, P.J., 1981. The preparation of thin sections of soils and unconsolidated sediments. Harpendon: Rothamsted Experimental Station.

Crowther, D.R., French, C.A.I. and Pryor, F.M.M., 1985. Approaching the Fens the flexible way. In: C. Hazelgrove, M. Millet and I. Smith (eds.), Archaeology from the ploughsoil, 59-76.

Catt, J.A., 1978. The contribution of loess to soils in lowland Britain. In: S. Limbrey and J.G. Evans (eds.), The effect of man on the landscape: the Lowland Zone. CBA Research Report 21, 12-20.

Evans, C. and Hodder, I., 1983/4. Excavations at Haddenham. Fenland Research 1, 32-36.

Evans, C. and Hodder, I., 1984/5. The Haddenham Project. Fenland Research 2, 18-23.

Evans, J.G., 1972. Land Snails in Archaeology. London: Seminar Press.

French, C.A.I., 1983. An environmental study of the soils, sediments and molluscan evidence associated with prehistoric monuments on river terrace gravels in northwest Cambridgeshire. University of London: Ph.D. thesis.

French, C.A.I., 1984/5. A Summary of Soil Analyses from the 1982/3 Excavations at Haddenham. Fenland Research 2, 23-25.

French, C.A.I. and Taylor, T., 1985. Desiccation and destruction: the immediate effects of de-watering at Etton, Cambridgeshire. Oxford Journal of Archaeology 4, 139-155.

Godwin, H., 1975. The history of the British flora (2nd edition). Cambridge: University Press.

Kwaad, F.J.P.M. and Mücher, H.J., 1979. The formation and evolution of colluvium on arable land in northern Luxembourg. Geoderma 22, 173-192.

Pryor, F.M.M., 1985a. Dyke Survey: An imperfect approach to the invisible. Archaeological Review from Cambridge 4:1, 5-14.

Pryor, F.M.M., 1985b. Current Archaeology 96, 6-8.

Pryor, F.M.M. and French, C.A.I., 1985. Archaeology and Environment in the Lower Welland Valley. East Anglian Archaeology 27. Fenland Project Monograph I.

Pryor, F.M.M., French, C.A.I. and Taylor, M., 1985. An interim report on excavations at Etton, Maxey, Cambridgeshire, 1982-84. Antiquaries Journal 65: 2, 275-311.

Macphail, R.I., 1987. A review of soil science in archaeology. In: H.C.M. Keeley (ed.), Environmental archaeology: a regional review 2, 332-379 HBMC. Occ. Paper.

R.C.H.M., 1960. A matter of time: an archaeological survey. Royal Commission on Historical Monuments (England).

Scaife, R.G., 1983. An interim report on the pollen analysis of Etton Neolithic causewayed enclosure. Ancient Monuments Lab. Report No. 3942.

Scaife, R.G., 1984. Further palynological data from Etton Neolithic causewayed Enclosure. Ancient Monuments Lab. Report No. 4396.

Shackley, M.L., 1975. Archaeological Sediments: A Survey of Analytical Methods. London: Butterworths.

Weir, A.H., Catt, J.A. and Madgett, P.A., 1971. Postglacial soil formation in the loess of Pegwell Bay, Kent (England). Geoderma 5, 131-149.

9. SO-CALLED SECONDARY PODZOLIZATION IN BARROWS

L.T. Runia

Abstract

Soils buried under barrows may be useful in the debate on human influence on podzol formation. However, buried soils may have been changed after being covered: possibly, a podzol profile may have been enhanced or even formed under barrows by a process called secondary podzolization.

Five barrows from the Netherlands were studied. Soil samples were analysed for the fractions of iron, aluminium and carbon involved in the podzolization process. Chemical data are discussed. The data corroborate current ideas on podzol formation. In the barrows studied, the buried soil profiles to a large extent show the soil profiles present during raising of the barrows.

9.1 Introduction

9.1.1 Are podzols man-made?

Podzols in NW Europe, in a sense, may be called man-made soils. However, human involvement in podzol formation is of a different character from the agricultural practices leading to man-made soils sensu stricto, for instance plaggen soils. Plaggen soils are connected to podzols by the agricultural system in which large areas of heathlands on podzols were used for sheep grazing and sod cutting, thus concentrating 'fertility' from large stretches of poor soils (podzols) on the relatively small area of the 'essen' (Pape, 1970). With the exception of the oldest plaggen soils, the presence of podzols in sandy parent material with low water holding capacities and, due to compact Bs and Bhs horizons, little rooting space appears to have been a prerequisite for the initiation of plaggen manuring (Heidinga 1988). By this system, man probably prevented reafforestation of the heaths, and thus to an unknown extent may have made continuation of podzolization possible. Man is also blamed for initial formation of pozols in the Neolithic/Bronze Age, when farmers destroyed primeval forests for arable land, a process that has been widely discussed (Limbrey 1975; Ball 1975; Casparie and Groenman-van Waateringe 1980).

9.1.2 Barrows and soil formation

In NW Europe, the soil profiles found in and below barrows and other prehistoric features played a role in the discussion on podzol formation and in the related discussion on human influences on spreading of heath vegetation in the Neolithic and Bronze Age. The general observation that heath vegetation and podzols occur jointly resulted in a model in which a causal relationship between these elements was assumed, the heath vegetation being responsible for the podzolization and the podzols being too poor to support anything but a heath vegetation (Edelman 1960; Havinga 1963).

Already in the 19th century, prehistoric features were used to date the podzolization process: Beyerinck (1933, 284) cites P.E. Müller (1887) who states that "die typische Haidebildung, mit Torf, Bleisand und Ortstein schon zu der Zeit, als die Hunengräber aufgeworfen wurden, voll entwickelt war, denn dieselben sind auf dem genannten Schichten aufgefürt".

A distinct role in the discussion on podzol formation in the Netherlands was played by the archaeologist van Giffen. Not trained in soil science, in some of the first of his many barrow excavations in the first quarter of the century, he mistook the extremely bleached podzol-E horizon under the barrow for a floor of white sand spread out for ritual reasons before raising of the barrow (fig. 9.1, van Giffen 1931). Later on, van Giffen not only understood the true nature of this layer of white sand, but also, thanks to his observations in many barrow excavations, made an important contribution to a symposium on podzol formation, held in 1941. This symposium was held because of the conflicting opinions on podzol formation. The then current ideas of podzolization by means of eluviation and illuviation (van Baren 1927) were challenged by a model based on palynological investigations of podzol profiles: Beyerinck (1933) stated that podzol B horizons were formed at the soil surface in cold periods under tundra vegetation, after which these 'tundrabanks' were gradually covered by a layer of bleached sand (fig. 9.2).

At this symposium, van Giffen showed that podzols were formed from the late Neolithic/Bronze Age onwards. In heathlands on podzols, the only non-podzol soils were the subsoils of Neolithic barrows, the podzol 'ruban tricolore' running in the top of these barrows and both underneath and over Bronze Age and younger barrows (van Giffen 1941). Since then, the presence or absence of a podzol profile has been one of the criteria in distinguishing Neolithic and Bronze Age barrows.

Besides their obvious archaeological importance, barrows remain of interest because of the preserved soils and pollen (Casparie and Groenman-van Waateringe 1980).

9.2 Secondary podzolization in barrows

9.2.1 The introduction and the concept of secondary podzolization

In 1963-64, two papers were published in which the soil profiles and the soil formation in and below barrows situated on poor cover sands in the Netherlands and Belgium were discussed (Scheys 1963; Waterbolk 1964). It was concluded, that soil forming processes might have changed the characteristics of soil profiles buried under barrows, in a sense that podzols below barrows might have been enhanced or even

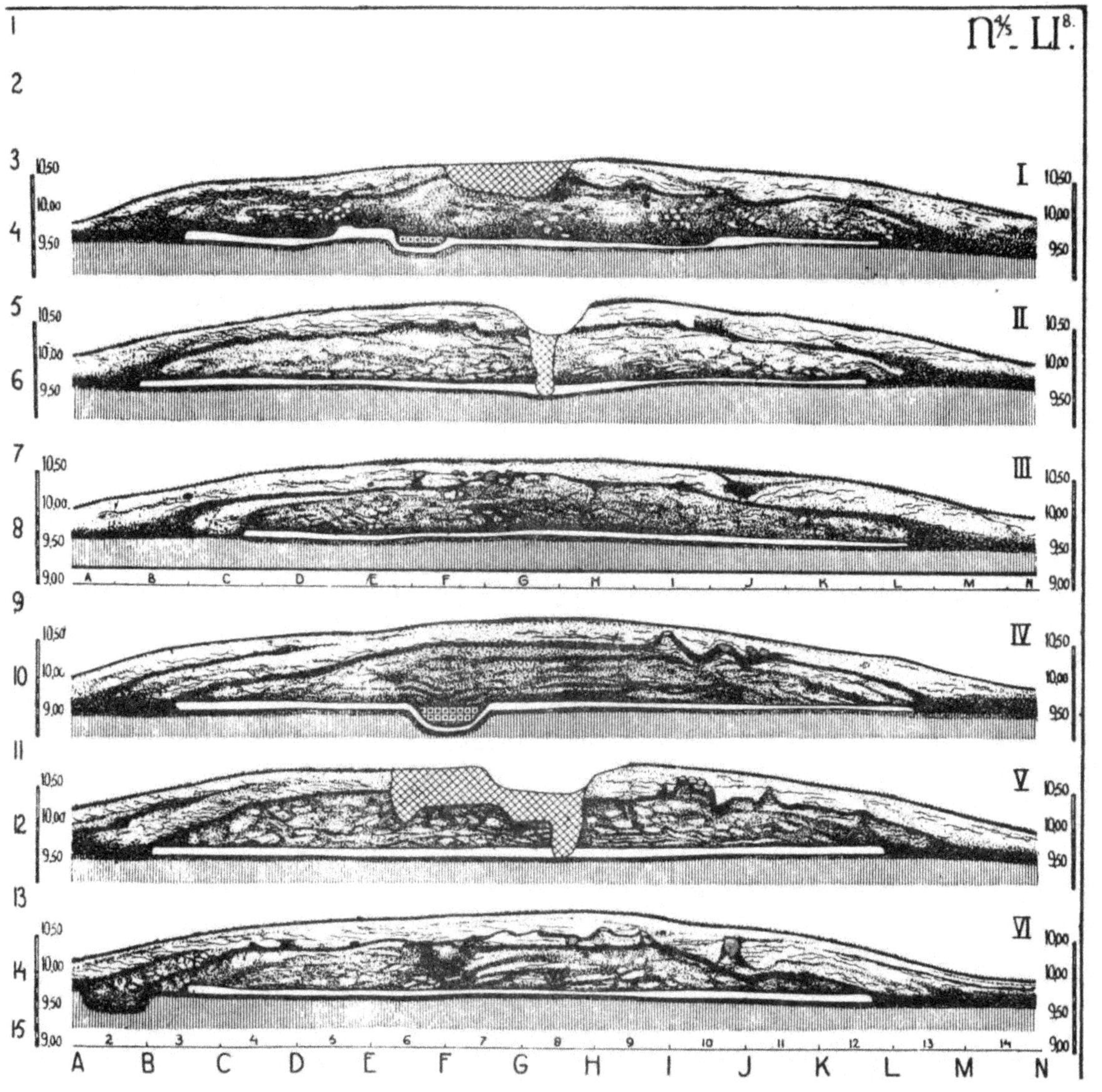

Fig. 9.1 Sections through an early Bronze Age barrow near Zeyen, prov. of Drenthe, excavated in 1919. This was one of the early barrow excavations of van Giffen, and he mistook the white buried E horizon for some kind of ritual floor of white sand. A podzol profile around the burial pit makes this barrow an example of a secondarily podzolized barrow (Drawing from van Giffen 1931).

Fig. 9.2 Schematic representation of the stratigraphy of the coversands, according to Beyerinck (1933). The dark podzol-B horizons are called "tundrabanks" and formed during, and correspond with, cold periods in the Wurm glaciation (Weichselian) (Drawing from Beyerinck 1933).

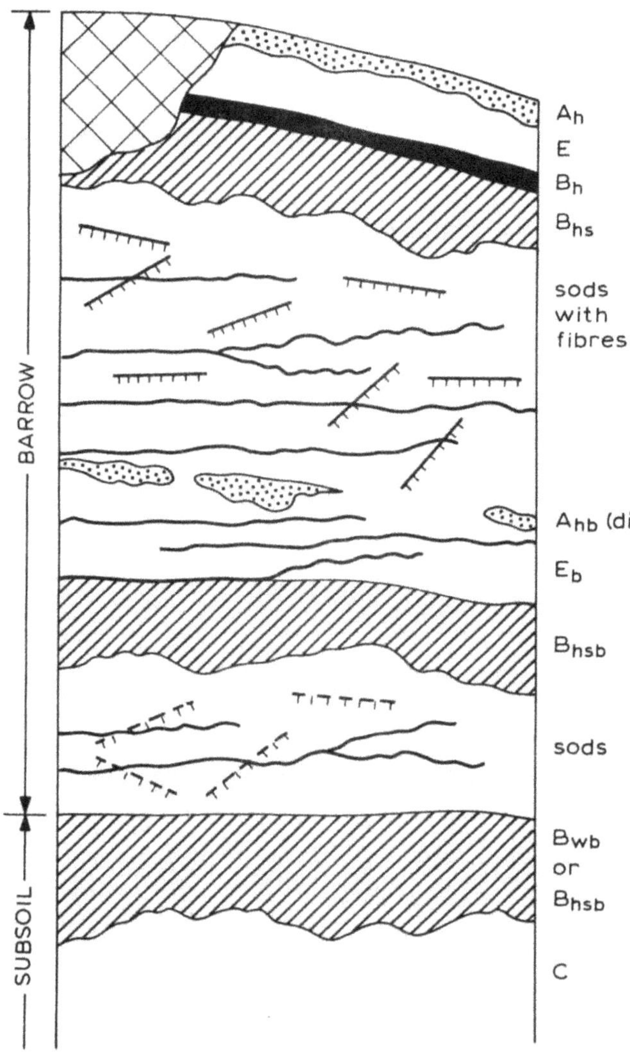

Fig. 9.3 Schematic section through a barrow. Horizon designations according to FAO 1974. Drawing: IPP.

formed by a process called secondary podzolization. This process obviously would influence our picture of the soils and of the landscape in which prehistoric man lived and would also have consequences for the discussion on man's role in causing podzolization.

According to Waterbolk (1964), characteristics of secondary podzolized barrows may include a grey coloured sod core, a bifurcated E (eluvial) horizon in the flank of the barrow, a sharp upper boundary of the buried B (illuvial) horizon and podzol profiles around postholes and burial pits below the barrow (fig. 9.1), which are the result of 'lixiviation' of the sod core of the barrow.

After 1964, pedogenesis in and below barrows received little or no interest from soil scientists, though discussions among archaeologists continued (Modderman 1975). The lack of pedological studies of barrows urged J.A. Bakker to initiate the investigation discussed briefly in this and more extensively in another paper (Runia and Buurman 1987).

9.2.2 Some chemical data from Dutch barrows

It was decided to study some barrows in relation to recent concepts on podzol formation (Petersen 1976; Mokma and Buurman 1982). Podzols are characterized by the presence of an eluvial and an illuvial horizon. In the upper horizon, Ah, organic matter is decomposed biologically and mixed with mineral soil material. Biological decomposition leads to the formation of large organic molecules, e.g. fulvic acids, that may become soluble due to uptake of limited amounts of the sesquioxides iron and aluminium. The organo-metallic complexes are soluble at carbon/sesquioxide (=Fe+Al) ratios above a critical level. The organo-metallic complexes formed in the upper part of the profile are transported down through the eluvial (E) horizon, which already has been depleted of Fe and Al, and precipitate in the illuvial horizon (B) as the carbon/sesquioxide ration drops below the critical level due to uptake of more Fe and Al. In well-drained podzols, below the B horizon lamellae (fibres), consisting of illuviated organic matter and sesquioxides, are present.

The fractions of iron, aluminium and carbon involved in podzolization can be determined using a pyrophosphate extraction (Bascomb 1968; Petersen 1976; Higashi et al. 1981; Mokma and Buurman 1982). The podzolization process is then characterized by the atomic ratio of pyrophosphate extractable carbon/sesquioxides (Cp/sesp), which measures the solubility of the organo-metallic complexes; the Cp/sesp ratio decreasing with depth. The critical value, below which the complexes precipitate, is found in the upper part of the Bhs horizon.

The deposition of carbon and sesquioxides in the fibres is controlled by physical rather than by chemical processes, as is reflected by the anomalous Cp/sesp ratios in fibres (Mokma and Buurman 1982). Probably low molecular organic compounds are responsible for the fibre formation.

Five barrows on sandy parent material were studied. One barrow (Eext) was sampled during excavation. Four barrows, which had been excavated previously but in which sections were still intact were selected for sampling. Criteria for selecting barrows were the presence of podzol profiles and possible secondary podzolization. Data concerning barrows are given in Table 9.1.

Soil samples were taken from various horizons (fig. 9.3) of the selected barrows, dried and sieved and treated with a 0.1M sodium pyrophosphate extraction solution. Carbon, iron and aluminium concentrations were measured in the pyrophosphate solution using atomic absorption spectrometry.

Atomic carbon/sesquioxide ratios (Cp/sesp) were calculated, this ratio being a measure for the stability of the complexes. These ratios were plotted against Cp, which measures the accumulation of complexes. Points representing similar horizons appear clustered in the plot (fig. 9.4).

Recent E and B horizons are separated by a Cp/sesp value of approximately 18. This agrees well with data from other well-developed podzols (Mokma and Buurman, 1982).

For our purpose, the Cp/sesp ratios of the buried E and Bhs horizons are of special interest. As with recent E and B horizons, here too the clusters are separated by a horizontal line in fig. 9.4 at Cp/sesp of ca. 9. As it is not likely that Cp/sesp values decreased after raising of the barrow (Runia and Buurman, 1987), we have to accept formation of the new buried podzols at relatively low Cp/sesp levels. This possibly indicates podzol formation at relatively high pH levels, as the critical Cp/sesp value appears to be pH dependent (Buurman 1985). From data of Mokma and Buurman (1982, appendix 4a), Cp/sesp values range from ca. 10 at pH=4 and ca. 20 at pH=3. Thus, from these data an acidification of soils since the Neolithic/Bronze Age might be inferred.

The Cp/sesp ratios in sods of the barrow studied are below the critical value of the present-day podzols, indicating that organo-metallic complexes formed in the top of the barrow cannot reach the buried soil profile. On the other hand, sods may be the cause of secondary podzolization (Waterbolk 1964). However, sods containing relatively large amounts of iron and aluminium, as in those taken from brown soils (Cambic Arenosols), are very unlikely to produce soluble organo-metallic complexes that can cause podzolization of the subsoil. Sods taken from already podzolized soils containing relatively large amounts of organic material and low amounts of Fe and Al, may be permeable for soluble compounds formed at the surface of the barrow. On decomposition of organic matter they may produce complexes that are soluble, reach the subsoil and enhance the already existing podzol, and eventually may cause the formation of <u>podsolsäume</u> around burial pits and postholes. Besides this, the organic matter in the sods may be the cause of the formation of fibres: the circumstances for biological decomposition in the sod core of barrows may be such, that low molecular organic compounds, which may be responsible for fibre formation, are formed.

As already stated by Modderman (1975), fibres are responsible for a number of secondary podzolization features, including the sharp upper boundary of buried B horizons (Plate 9.1). Our data corroborate

Table 9.1 Some characteristics of the barrows studied.

No. of barrow	No. of periods	Date of first period	Parent material	Soil profile ** below barrow	Soil profile on top ** and around barrow	Reference
1. Eext	2	3450–3200 bp	fine sand (coversand)	cambic arenosol	orthic podzol	Lanting 1984
2. Meelworstenberg	2	3450–3250 bp	medium sand (sand deposit)	orthic podzol	orthic podzol	Glasbergen et al 1947–48; Glazema 1951
3. Maarsbergen I	2	Late Neolithic*	fine sand (coversand)	orthic/leptic podzol	orthic podzol	Lanting and van der Waals 1971
4. Swalmen HT 8	2 or 3	3305 ± 30 bp (GrN 6395)	loamy coarse sand (Rhine terrace)	cambic arenosol	orthic podzol	Lanting and van der Waals 1974
5. Anner Tol III	4	3870 ± 35 bp (GrN 6634)	fine sane (coversand)	orthic/leptic podzol	orthic podzol	Butler et al 1972

* Early Bronze Age, c. 3650–3450 bp (Lanting and Mook 1977)

Late Neolithic, c. 3800 bp (Lanting and Mook 1977)

** FAO, 1974

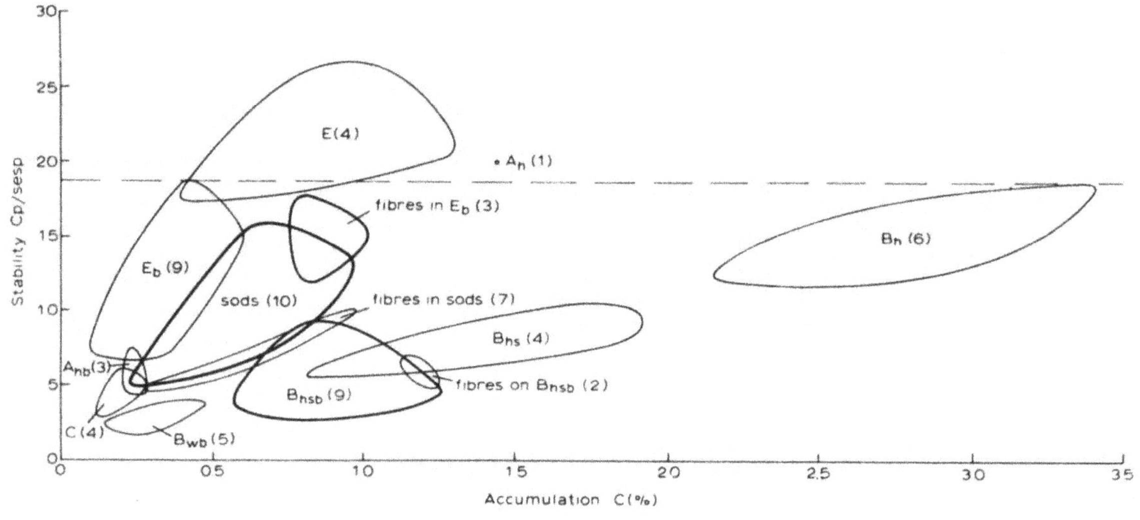

Fig. 9.4 Clustering of data points. On the horizontal axis, the accumulation of organo-metallic complexes, as measured by Cp (mass %) is given. On the vertical axis, the stability of the complexes, as measured by the atomic ratio Cp/sesp is plotted. Stability of the complexes decreases with increasing Cp/sesp. In brackets the number of points in each cluster. Horizon designations correspond to those given in fig. 9.3. Drawing: IPP.

Plate 9.I Original soil profile (Eb and Bhsb) buried under the barrow 'Meelworstenberg'. The very white Eb horizon contains fibres (lamellae), and also on top of the Bhsb a fibre is visible. Scale in cm. Photo: IPP.

current ideas on fibre formation: fibres have Cp/sesp ratios comparable to those of the horizons in which they are found (fig. 9.4), and besides this, in barrows fibres are usually found in Eb and on Bhsb horizons (Plate 9.1), locations explainable by physical mechanisms (stopping waterfront, filtering).

Another characteristic of secondary podzolized barrows, the bifurcated E horizon in the flank of a barrow, may be explained by the chemical behaviour of the organo-metallic complexes. It is obvious that the complexes will not precipitate in existing E horizons, because the virtual absence of sesquioxides prohibits decrease of the Cp/sesp levels to below the critical value. This means that the Bhs horizon of the surface podzol will 'jump over' the original E horizon.

9.3 Conclusions

In spite of the relatively small number of barrows investigated, of which none had a podzolized grave pit, results of the chemical analysis are of interest. In the barrows studied, it is unlikely that buried soil profiles act as illuvial horizons for organo-metallic complexes formed at the present-day soil surface. Though it is very unlikely for podzol profiles to be formed below barrows, existing podzols may be enhanced to a certain extent. Apart from the fibres, the profiles buried under the barrows investigated to a large extent show the soils present during the period in which the barrows were raised. The characteristics of secondary podzolization may be explained largely by normal podzolization processes. However, investigation of a barrow with a podzol profile around the grave pit remains to be done.

Our data suggest a decrease of soil pH since the raising of the barrows in the Neolithic and Bronze Age.

Acknowledgements

Special thanks are due to Dr. J.A. Bakker, Dr. P. Buurman, Professor J.D. van der Waals, Professor H.T. Waterbolk and Drs. J.N. Lanting. Drawings were made by B. Donker (IPP) and photos by F. Gijbels (IPP).

References

Ball, D.F., 1975. Processes of soil degradation: a pedological point of view. In: J.G. Evans, S. Limbrey and H. Cleere (eds.), The Effect of Man on the Landscape: The Highland Zone. CBA Research Report 11, 20-27.

Baren, J. van, 1927. De Bodem van Nederland, deel II. Amsterdam: van Looy.

Bascomb, C.L., 1968. Distribution of pyrophosphate-extractable iron and organic carbon in soils of various groups. Journal of Soil Science 19, 251-268.

Beyerinck, W., 1933. De oorsprong onzer heidevelden. Nederlands Kruidkundig Archief 43, 272-292.

Butler, J.J., Lanting, J.N., and van der Waals, J.D., 1972. Anner Tol III: A four-period Bell Beaker and Bronze Age barrow at Schuilingsoord, gem. Zuidlaren, Drente. Helinium 12, 225-241.

Buurman, P., 1985. Carbon/sesquioxide ratios in organic complexes and the transition albic/spodic horizon. Journal of Soil Science 36, 255-260.

Casparie, W.A. and Groenman-van Waateringe, W., 1980. Palynological analysis of Dutch barrows. Palaeohistoria 22, 7-65.

Edelman, C.H., 1960. Podzols forestières et podzols de bruyère. Pédologie 10, 229-249.

Food and Agricultural Organization, 1974. Soil Map of the World. Vol. I Legend. Paris: Unesco.

Giffen, A.E. van, 1931. Die Bauart der Einzelgräber. Leipzig: Mannus-Bibliothek 44-45.

Giffen, A.E. van, 1941. De tijd van vorming van heidepodsolprofielen aan de hand van archaeologische waarnemingen. In: Besprekingen over het heidepodsolprofiel. Arnhem: Koninklijke Nederlandse Heidemij, 12-23.

Glasbergen, W., Praamstra, H. and Waterbolk, H.T., 1947-48. Unpublished excavation reports, Biological-Archaeological Institute, Groningen.

Glazema, P., 1951. Oudheidkundige onderzoekingen in de provincie Gelderland. Gelre 51, 89-99.

Havinga, A.J., 1963. A palynological investigation of soil profiles developed in cover sand. Mededelingen van de Landbouwhogeschool te Wageningen, Nederland 63 (1), 1-93.

Heidinga, H.A., 1988. This volume.

Higashi, T., De Coninck, F. and Gelaude, F., 1981. Characterization of some spodic horizons of the Campine (Belgium) with dithionite-citrate, pyrophosphate and sodium hydroxide-tetraborate. Geoderma 25, 131-142.

Lanting, J.N., 1984. Kroniek van opgravingen en vondsten in Drenthe in 1982. Nieuwe Drentse Volksalmanak 101, 61-63.

Lanting, J.N. and Mook, W.G., 1977. The Pre- and Protohistorie of the Netherlands in terms of radiocarbon dates. Groningen.

Lanting, J.N. and van der Waals, J.D., 1971. Laat-neolithische grafheuvels bij Vaassen en Maarsbergen. Oudheidkundige Mededelingen Rijksmuseum Oudheden Leiden 52, 93-127.

Lanting, J.N. and van der Waals, J.D., 1974. Oudheidkundig onderzoek bij Swalmen. I. Praehistorie. Oudheidkundige Mededelingen Rijksmuseum Oudheden Leiden 55, 1-111.

Limbrey, S., 1975. Soil Science and Archaeology. London: Academic Press.

Modderman, P.J.R., 1975. Bodemvorming in grafheuvels. Analecta Praehistorica Leidensia 8, 11-21.

Mokma, D.L. and Buurman, P., 1982. Podzols and podzolization in temperate regions. ISM Monograph 1. Wageningen: International Soil Museum.

Pape, J.C., 1970. Plaggen soils in the Netherlands. Geoderma 4, 229-256.

Petersen, L., 1976. Podzols and podzolization. Copenhagen: DSR Forlag.

Runia, L.T. and Buurman, P., 1987. The so-called 'Sekundärpodsolierung' in burial mounds: chemical data from Dutch barrows. Journal of Archaeological Science 14, 97-105.

Scheys, G., 1963. Podzolvorming belicht door archeologische waarnemingen. Pédologie 13, 216-230.

Waterbolk, H.T., 1964. Podsolierungserscheinungen bei Grabhügeln. Palaeohistoria 10, 87-102.

10. MEGALITHS, SOILS AND VEGETATION ON THE DRENTHE PLATEAU

J.A. Bakker and W. Groenman-van Waateringe

Abstract

The distribution and site location of the Neolithic megalithic tombs (hunebedden) on the Drenthe Plateau (fig. 10.1) were analysed. The recent destruction of these tombs seems to have been random without affecting the original pattern too much (section 2).

Because all hunebeds were used during TRB pottery phase D, Thiessen polygons around them (fig. 10.2) would, according to Renfrew, show territories of which the tombs were the central marker. First of all, however, they demonstrate a very unequal tomb distribution (section 3), which is due to the unequal availability of boulders (section 4). Boulders were available in areas where Pleistocene erosion had delivered them from the encasing boulder clay (fig. 10.4). Most hunebedden lie less than 350m from such places according to the 1:50,000 geological and soil maps.

Many tombs were situated on routes of communication, a more promising object of study than polygon territories (sections 5-6, figs. 10.5-6).

The Drenthe Plateau is covered by 'coversand' which is usually loamy and thinner than 1.2m when lying on top of boulderclay. Otherwise the sand is loam-free. Such loamless sands with a low ground water table were preferred for hunebed location, and most tombs would have been good vantage points if we disregard the limits of the contemporaneous forest clearings (section 7, figs. 10.7-8).

Palynological analysis of the old ground surface below hunebed barrows points to a correlation between a short distance to loamy coversands (code 23), low *Tilia* values, relatively low AP values and relatively high values for Cerealia plus *Plantago lanceolata* on the one hand; and on the other hand, a distance of over 100m to code 23 soils, high *Tilia* values, high AP values and low values for Cerealia plus *Plantage lanceolata*. This is explained by assuming that code 23 soils, on which the more demanding *Tila* and *Fraxinus* grew, were preferentially cleared for arable (section 8, figs. 10.9-13).

10.1 Introduction

The Neolithic Funnel Beaker culture (TRB, 4,700-4,100 bp in conventional radiocarbon years) is found in the Netherlands almost exclusively on the pleistocene sandy soils north of the Rhine.

Virtually no remains have been recovered in the 40-90km wide wetland zone along the coast. This has unfortunate consequences for the researcher. Whereas, optimistically, we may reckon to know almost every detail of the ecology, the landscape, the food and the perishable organic artefacts of wetland cultures like the contemporary Vlaardingen culture and the Swifterbant culture of six hundred years earlier, or from wetland sites of the later Corded Ware culture, similar data from TRB in Holland are nonexistent or extremely rare. Everything organic and unburnt has perished in the dry, non-calciferous and now acid sandy soils.

Even soil discolourations, the pride and the working principle of the excavator in such podzol bearing sands, are denied to us for TRB. For at this time, podzolization had scarcely, if at all, started on the high sands where TRB remains are found and where these first known farmers of the region cleared the primeval forests. Podzolization only became a general phenomenon with the subsequent expansion of grass- and heathlands from Bell Beaker times and Bronze Age onwards.

On the other hand the distribution pattern of TRB is well known especially in Drenthe[1], where besides the obvious megalithic graves, which are called hunebedden, many sherds were collected by amateur archaeologists from heathlands in reclamation and from the present arable fields. Distributions of wetland sites are often determined by the lay-out of levees, dunes, lakes and the like, and may then show linear arrangements; but for TRB on the former heathlands this is much less so and one might even suppose that social factors were the sole determinants for the distribution of its settlements, hunebedden and inhumations without above-ground monuments.

At present three parallel series of very detailed geo-maps on sheets of the 1:50,000 topographical map are in the course of publication for the region: a very rare luxury for the archaeologist. The Geological, the Geomorphological and the Soil Map are independently surveyed and published. By using these maps, TRB distribution patterns on the Drenthe Plateau in the northeast of the country are analysed and discussed in sections 2-7 by the first author. He also investigates the applicability for hunebeds of Colin Renfrew's concept of megalithic graves as central places in Thiessen polygon territories[2]. Then the second author reinterprets the available pollen spectra from old land surfaces below primary barrows of hunebeds (section 8).

The Drenthe Plateau is a low and very flat, sand covered, ground moraine plateau traversed by a number of parallel river valleys of which the province of Drenthe is the core, but which includes margins of the adjoining provinces of Groningen, Friesland and Overijssel in the northeast of the country (fig. 10.1). Its highest part is in the southeast and does not quite reach 30m above sea level (30m + NAP)[3]. Its lowest parts are in the west and north, where peat and clay layers cover its surface at about NAP.

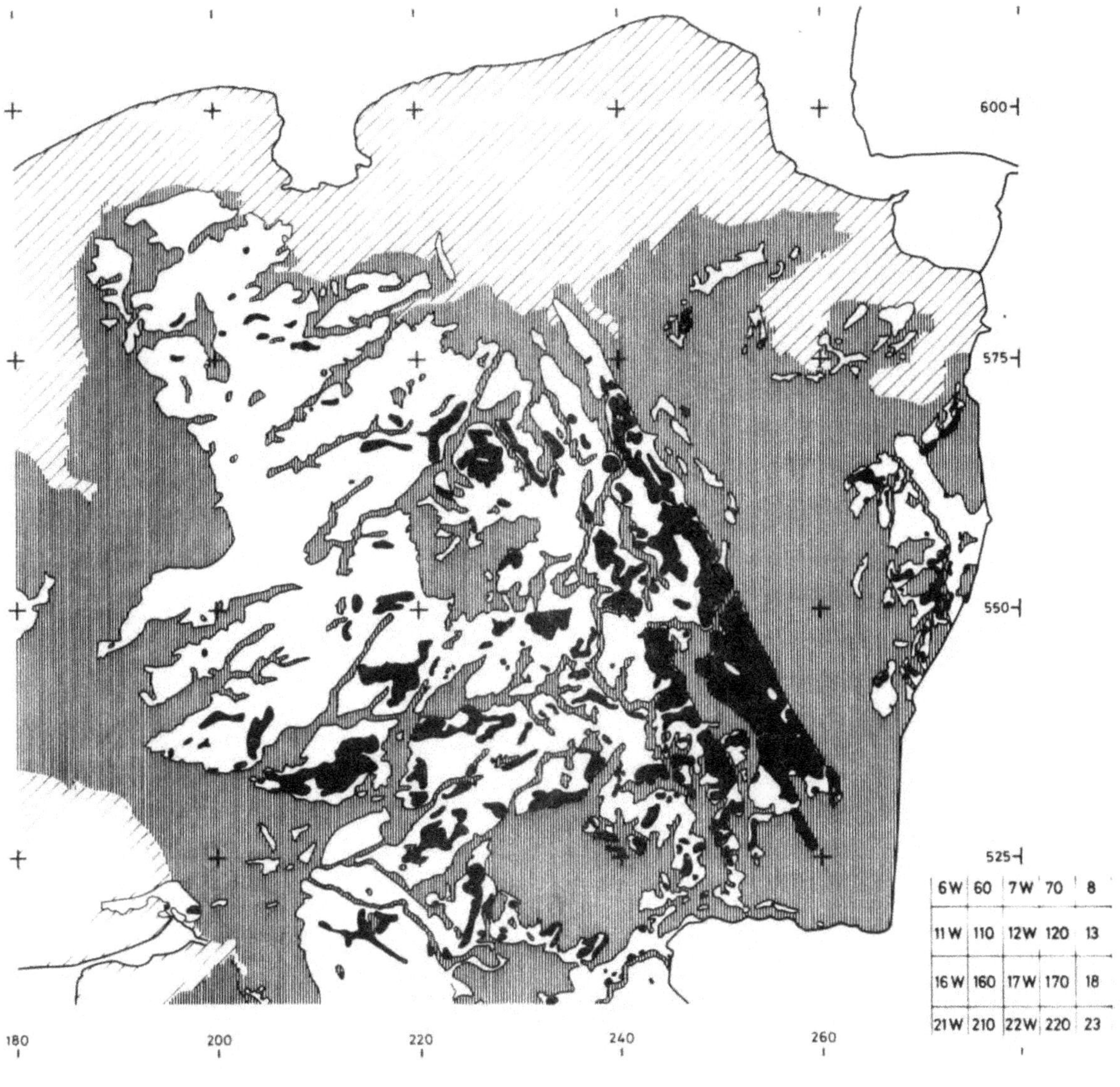

Fig. 10.1 The sandy Drenthe Plateau (black and white) surrounded by peaty soils (hatched vertically) and clay (obliquely hatched). Mean winter ground water table is deeper than 1m in black areas and higher in blank areas.
The E-W distance of the crosses is 20km, the N-S distance is 25km; they indicate the corners of the sheets of the 1:50,000 maps (inset). Source: Bakker 1980, based on the 1:200,000 COLN and NEBO maps. Drawing: IPP.

10.2 Uncertainties about the former distribution of hunebeds and boulders

The distribution of hunebeds is dependent on the availability of the large blocks from which they were constructed, but the original frequency of erratic blocks at the surface of the once glacier-covered TRB West group area is not known, because the country was stripped of its boulders in historical times. Nor has a map yet been compiled of the size and frequency of recently found blocks like the one developed for the German Democratic Republic by W. Schulz and other researchers (Kahlke 1981, ill. on p. 60; Gehl in Schuldt 1972, map 13). Geologists tell us that boulders were more common on the Drenthe groundmoraine Plateau than in the terminal push moraine landscapes elsewhere in the Netherlands. Although there is much truth in this statement, one of their arguments is the presence of hunebeds in Drenthe and their absence elsewhere - which is not an independent answer to the question as to where TRB man could have built his megaliths.

Boulders have for long been a much sought after commodity. In the 11th to 13th centuries, they were gathered for foundations of churches and other large buildings. Between 1150 and 1250, boulders were cleaved into large cubic blocks for the masonry of churches. In the 18th and 19th centuries they were used much more intensively, being broken into pieces with the aid of blasting powder for dike revetments (which started with the pile worm catastrophe of 1731 and ended with the import of German basalt after 1860) and for metalling macadam roads (invented by John Loudon Mac Adam in 1819 and replaced by asphalt roads about 1880). Boulders are now collectors' pieces for gardeners, who often transport them over several kilometres.

Although the country is therefore almost bare of boulders in their original position on the surface, finds in sand and loam pits suggest that in TRB times boulders abounded on some parts of the Drenthe Plateau and were also present on the pleistocene push moraines of the other provinces north of the Rhine. The 1874 railway dam at Crailose Brug north of Hilversum, for instance, is lined by more than 100 large boulders, half of which are over 1m long, a quarter over 1.5m, and four blocks 2m long. The larger boulders would have sufficed for construction of a moderate tomb, if this collection were available before the excavation of a local till layer in 1874. There are, however, several 16th to 19th century reports about boulders lying at the surface near Hilversum until stone trade became good business in the 18th and 19th centuries.

Drenthe was one of the first legislatures in the world to protect its ancient monuments, acting in 1734/5 to safeguard the hunebeds. The rapid expansion of the pile worm, Teredo navalis, and its catastrophic destruction of all wooden dike fences in the salt and brackish waters of the Netherlands and Ostfriesland had prompted the invention of the modern stone covered dike. This started an active stone trade from the poor pleistocene areas of the Netherlands, northwestern Germany and Scandinavia, wherever transport was available to the endangered polders. Soon, the Drenthe authorities were compelled to prohibit the removal of boundary stones and the like. In the same long breath they enacted that "the so-called Hunebeds, which everywhere, as venerable monuments and time-honoured memorials, should be preserved" (Bakker 1979b).

In other provinces extending onto the Drenthe Plateau and no doubt also elsewhere in the northern and central Netherlands, those hunebeds which had survived claims of medieval architects disappeared rapidly. Thus the Eeze tomb (O1), some 75m across the Drenthe border in Overijssel, was blasted to pieces in the 1840s. A sketch from 1781, a depiction on an old estate map and a brief mention of its destruction documented its existence before this was established by the excavation in 1918 by Van Giffen (1927).

Many hunebeds must have disappeared completely unnoticed. In Westphalia, just across the Dutch border, hunebeds extended as far southward as the loess boundary in the Ruhrgebiet, which implies that they probably have ocurred throughout the whole Dutch TRB area, though more sparsely outside Drenthe (for a definition of the TRB area, see Bakker 1979a; 1982).

Even the legal protection in Drenthe did not preserve the remaining distribution pattern completely. Van Giffen (1925, 188) estimated that "originally there would have been at most double the present number" of hunebeds. Yet his estimate was based on the available documents from the late 17th till 19th century and he forgot that he borrowed the phrase from Johan Picardt (1660), who wrote in what now seems to have been a calm in the dilapidating activities. He guessed it "credible that half of them do not exist any longer, but are destroyed" (p. 131). In Picardt's days even the medieval cleaving technique with sledge-hammer and wedges was forgotten, because he described it as 'sawing'.

Van Giffen and successors also discovered several completely undocumented destroyed hunebeds and recently Huiskes (1985) traced in addition 96 potential (but as yet unproven) former hunebed sites on basis of such field names as Steenbergen, Steenakkers and Steenveen (Steen = stone; berg(en) = heap(s), mound(s); akker(s) = field(s); veen = peatbog).

The distribution pattern of known TRB megaliths on the Drenthe Plateau (fig. 10.2) must obviously have been much biased by subrecent destruction. Despite the destruction of several stone cists (reported by Tiesing (1974) and others), and some hunebeds, most of the sites of the latter are known and figure on the map. For instance the empty areas near grid points 252/547 and 248/555 (fig. 10.2) may have had one or more hunebeds, but the main impression of the distribution map seems not much distorted to us.

The distribution of the tombs is very uneven - concentrations occur on the straight easternmost Hondsrug and parallel ridges, but is much sparser west of them. In particular, the omission of some monuments which were destroyed early on will influence the pattern.

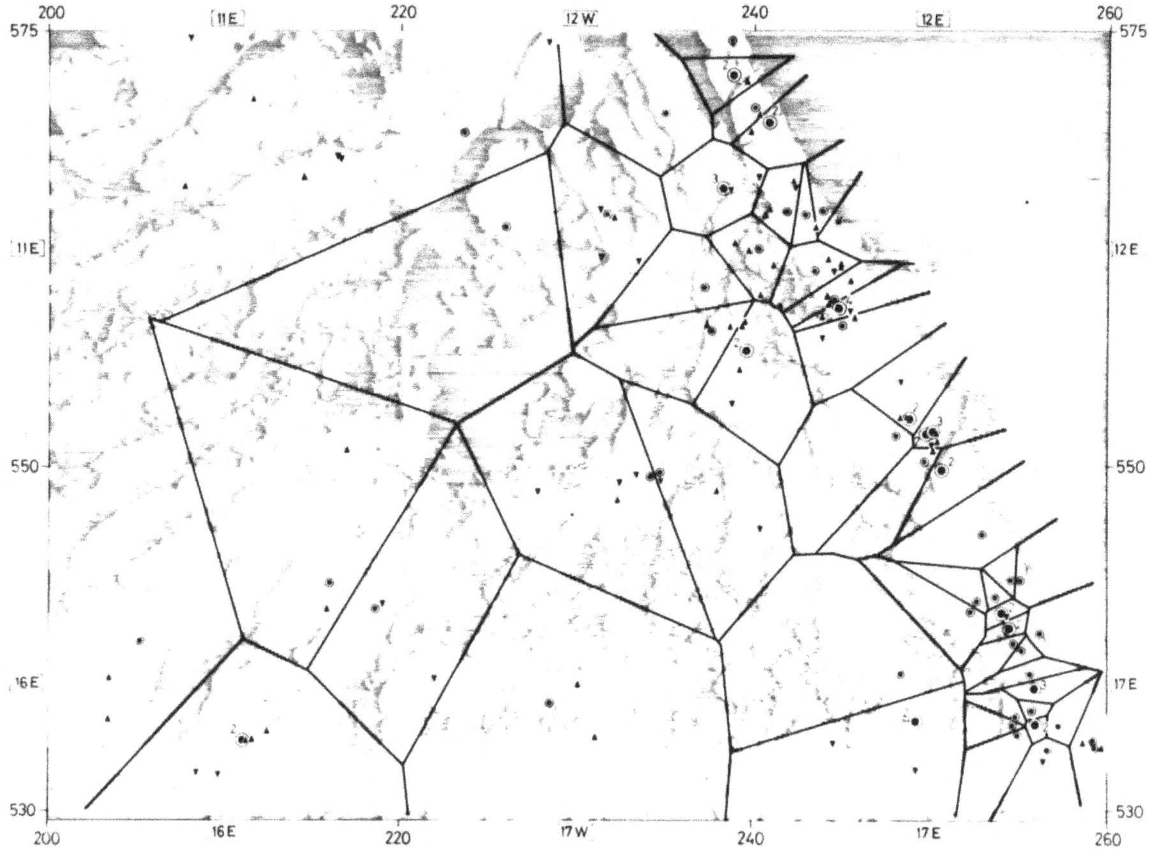

Fig. 10.2 Part of the Drenthe Plateau (map sheet numbers and grid references in margin, cf. fig. 1).
Thiessen polygons were constructed around the hunebeds (dots in a circle. Numbers indicate clusters, otherwise single).
Standing triangles: TRB settlements and single sherds.
Pendant triangles: TRB single earth graves and a stone cist. Small streams meandered through the (hatched) peaty periglacial valleys. Source: Bakker 1980. Drawing: IPP.

10.3 The Renfrew game

Starting from Childe's observation that in Scotland the megalithic tombs lie near the rare arable land and the fact that 17 of the 18 tombs on the isle of Arran, Scotland, are within 500m of land suitable for agriculture, Renfrew (1973a) drew Thiessen polygon 'territories' around the megaliths of Arran (fig. 10.3), the Orkney island of Rousay and the Maltese Islands. For Arran and Rousay, where arable soil is scarce and the tombs are evenly spread, this gave a not implausible division of the fields between the polygons. Since the mid-perpendiculars of a triangle all cross each other at a point, such polygons give a pleasing effect and a satisfaction to the draughtsman - and are spuriously over-persuasive.

As a cartographical experiment, the Renfrew polygon territories were drawn around the hunebeds on the Drenthe Plateau (fig. 10.2; Bakker 1980). An objection often heard to territory construction by Thiessen polygons around megaliths, namely that such maps are in reality palimpsests containing successive tombs all of which were never in contemporary use, does not hold for Drenthe in the middle of the TRB period. Some monuments were of course built earlier than others, but all contain Drouwen D pottery demonstrating simultaneous use.

But do these polygons reflect prehistoric reality in any way? What they show primarily is the unequal tomb distribution described above. Nevertheless in three areas regularities may be observed. In the southwest (grid 220/230 - 530/550) the tombs are spread thinly but evenly and the dividing lines coincide often with the peaty primeval valleys. Secondly, in a northern zone (220/240 - 560/575) six known hunebeds are neatly arranged 5.9 to 7.0km apart, on average 6.2km from each other. Thirdly, several megaliths of the Hondsrug are neatly aligned suggesting location on straight roads. These alignments which are very clear on 1:50,000 maps are often quite incompatible with any Thiessen polygon arrangements.

In a great number of publications Renfrew (1973a; 1973b; 1974; 1976) has argued that such megalithic tombs must have functioned as 'territorial markers' and were built by agriculturists marking and claiming their territories by such impressive religious monuments - not so very different from the visual effect achieved by our church steeples, which even gave rise to an 'esprit du clocher'. In Renfrew's maps the central place concept is taken literally - as would be the overall result of a similar analysis of the medieval churches of Drenthe lying in the centre of marke territories.

That the cartographical concept more or less worked for some landscapes with megalithic monuments was not contradicted by the situation mentioned for Rousay and Arran. Nevertheless most megalithic tombs of Britain and Ireland contain so very few pottery fragments, and typically decorated sherds are so extremely rare among these, that it seems almost out of the question ever to identify the connections between the (rarely known) settlements and the tombs, thus giving more body to the concept.

Superficially the situation in Drenthe seems to be much better.

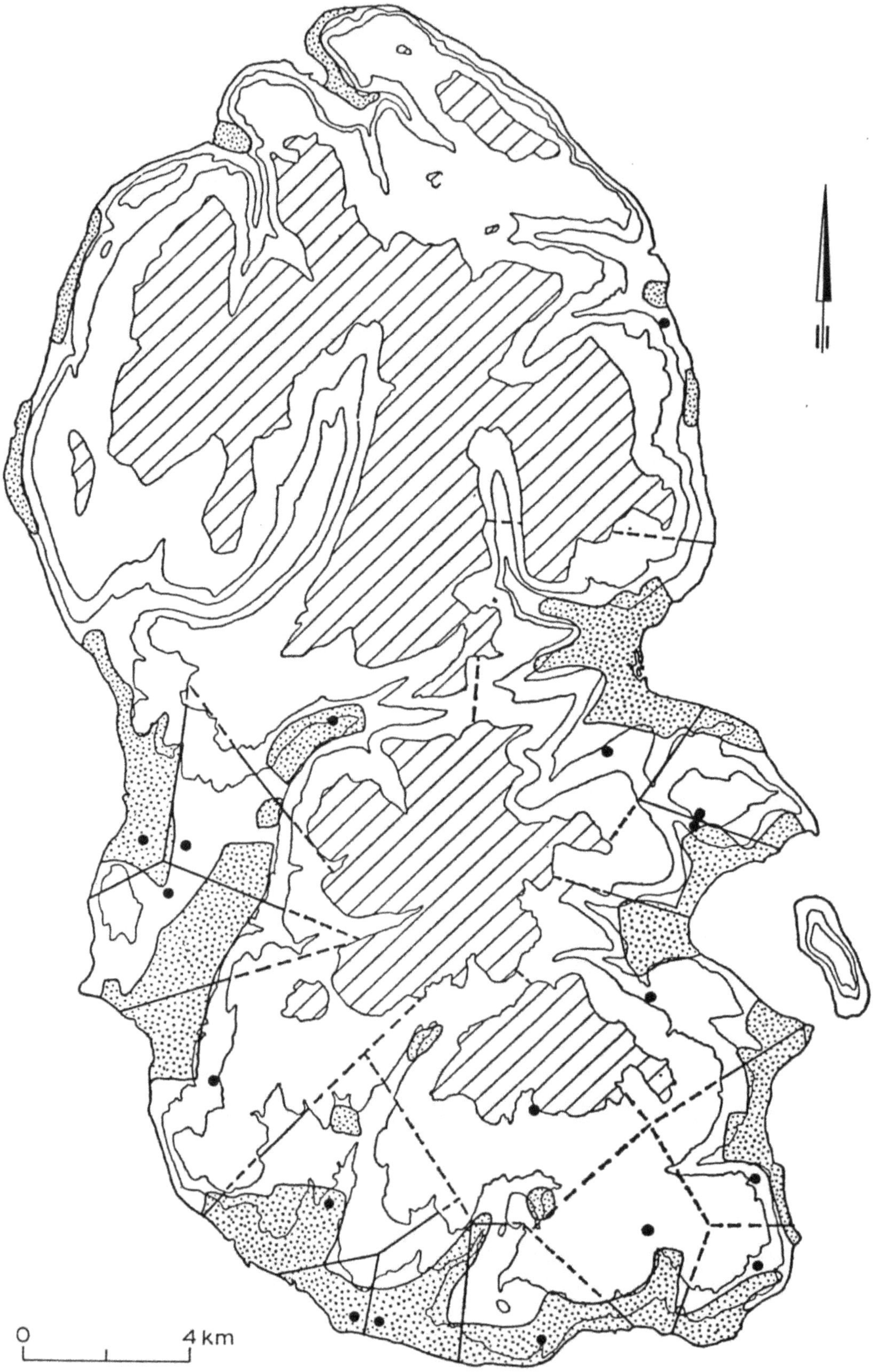

Fig. 10.3 Renfrew's Thiessen polygon territories around the chambered cairns of Arran.
Modern arable land is stippled, contours at 100m intervals.
Source: Renfrew 1973a.

The hunebeds deliver tens and hundreds of exquisitely decorated pots, most of which show a unique combination of motifs. Besides, private and museum collections contain many a decorated sherd from settlement sites now lying in arable land. Sherds from surface collections are, however, usually very small and the number of well decorated, first rate ritual ceramics from settlements is small. Trying to identify the domestic refuse of the potters which are known by the pecularities of their products in neighbouring graves is thus looking for white ravens - which, however, have not yet been found.

Within the chronological, regional and local varieties of style of the pottery from the excavated hunebeds, the products of individual potters are now beginning to be recognised (Luijten 1985).[4] Excavation of the newly discovered sacrifical plaza or parvis in front of some hunebed entrances, where a similar top quality pottery was deposited simultaneously with ceramics deposition in the chamber, seems also promising in this respect.

10.4 The availability of boulders

Let us now consider the geological basis for the availability of boulders to the hunebed builders on the Drenthe Plateau. Three factors are involved.

(a) These large blocks - of which those over 25 tons were not used for hunebeds - were brought here by the glaciers of the Saalian ice age. Although a small (but in fact unknown) number of boulders may have lain on the surface after the melting of the ice, the greater part by far was enclosed in the till or boulderclay layer covering the Plateau. In it the larger blocks were not evenly distributed; often they were packed in clusters.

The suggestion of an esteemed colleague that the TRB population dug large pits in the stiff till to find boulders is unacceptable - even with bulldozers and draglines this is an almost impossible job, also the distribution of boulders in the till is unpredictable.

(b) As soon as the ice had melted, erosion started. The present valley systems were cut into the plateau and the thickness of the till was reduced. Next to the deepest river valleys the till was completely removed leaving fields of washed out boulders, stones and pebbles. Such 'stonefields' - as I shall call them henceforth - are concentrated near the border of the Drenthe Plateau in the east and in the southwest (fig. 10.4). Erosion was strong here due to the deep valleys of the primeval Hunze directly east of the Hondsrug and of the primeval Vecht following the course of the Reest in the southwest. Many boulders must also have lain on top of the boulderclay in between the stonefields, because what was left of the till was very much worn off and reduced.

(c) After the warm Eemian period, the Netherlands were not covered by the Weichselian glaciers. In a polar desert sand storms covered the plateau with a blanket of 'coversand' layers varying in thickness from zero to over 3m in depressions. Where the coversand

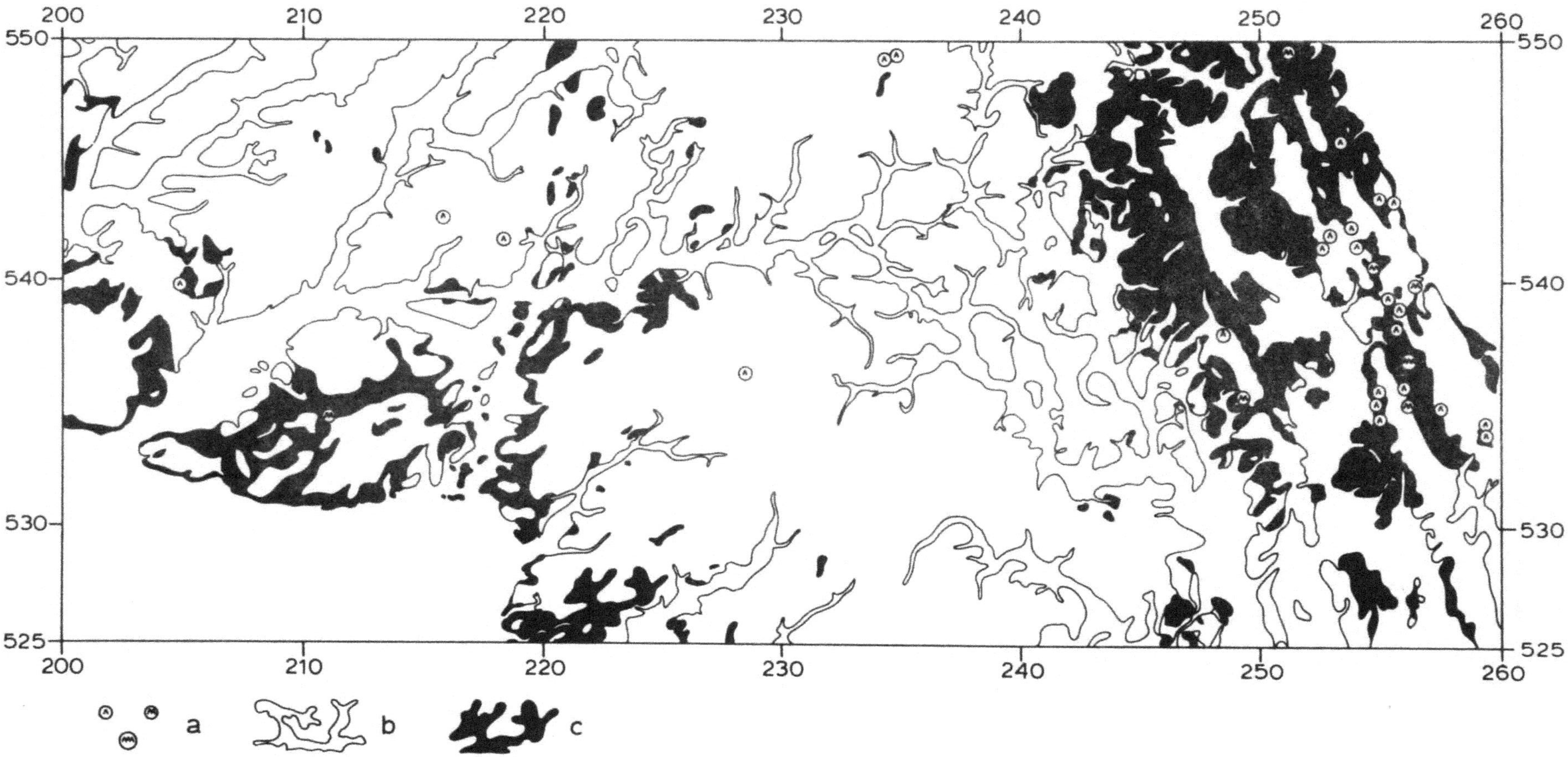

Fig. 10.4 Map of the southern Drenthe Plateau showing the close connection of the hunebed sites to areas where heavy periglacial erosion had washed boulders out of the groundmoraine till or boulderclay. a = 1-3 hunebeds. b = little to not completely eroded boulderclay. c = stony residuals of completely eroded boulderclay ('stonefields').
Source: Bakker 1982, based on the appendix maps of the Top of the Drente Formation of the Geological map sheets 16E, 17W, 17E. Drawing: IPP.

blanket was thin or absent tomb building material could be collected in the outwashed stonefields by TRB man. Fig. 10.4 shows the place of the stonefields, but not where the sand cover must have hidden even the stones that were over 1m thick. It should also be pointed out that the hunebeds are always located on dry sand, usually coversand of over 1.2m thickness. The boulders would therefore have been hauled some distance.

Of the 54 hunebeds still in existence (one was discovered in 1982) and the known sites of 24 destroyed ones in the Netherlands, respectively 53 and 21 lie on the Drenthe Plateau (fig. 10.2).[5]

Of these 74 tombs:

12 lie on sheet 12W (no geological or soil map published)
22 lie on sheet 12E (soil map published)
5 lie on sheet 16E (geological map published, concept soil map ready)
3 lie on sheet 17W (geological and soil map published)
32 lie on sheet 17E (geological and soil map published)

Only the recent 1:50,000 geological and soil map sheets are used here.

Fig. 10.3 was compiled from the appendix maps 'Top of the Drente Formation' of the geological map sheets 16E, 17W and 17E.

Of the 40 hunebeds on these sheets,

20 lie above completely eroded boulderclay ('stonefield')
20 lie above partially eroded boulderclay.

Of the latter 20 tombs,

1 has an unknown distance from stonefields (D44a)
4 are no more than 100m from a stonefield
4 are 150-200m from a stonefield
3 are 300m from a stonefield.

Thus in 78% of the known cases, the hunebeds were constructed less than 350m from areas where stonefields lay in the subsoil (or at the surface if there was a hole in the coversand blanket).

Of the remaining eight tombs the distance to stonefields is much larger and must the builders have resorted exclusively to boulders lying on top of the partially eroded boulderclay. From west to east these hunebeds are named D52a, D52, D54a, D54b, D54c, D32, D36, D37. The last three are on sheet 17E, on the Hondsrug. For D36 and D37 boulders must have been available in the nearby eroded river valley. The other five are from areas on sheets 16E and 17W where completely eroded boulderclay areas are rare. The contour lines on the geological appendix maps show that the tombs D52 and D52a also lie on or near the slopes of eroded river valleys. Where this seems not to be the case

(D54a, D54b-c with grid references 228/536 and 234/549, cf. fig. 10.4), two tombs with five pairs of side stones and one tomb with two pairs were built. They are not the shortest in their class (Bakker 1979a, 150, Table V), so that the expectation that such exceptionally located hunebeds were made only from small stones and had mediocre dimensions is not fulfilled.

The interim conclusion may be drawn that almost all hunebeds are restricted to areas where boulders were available according to the given interpretation of the geological map. Their uneven distribution is wholly due to the rarity of boulders (because of the lesser erosion of the boulderclay) in greater parts of the Drenthe Plateau.

The soil map tells us where the building materials for the hunebeds may be supposed to have been available at the surface. It documents boulderclay when thicker than 20cm and 0-40cm below the surface (symbol KX), or at 40-120cm below the surface (symbol x). The usual thickness of coversand on the Hondsrug where most hunebeds occur is 60-150cm. Thus most larger blocks on top of the partly eroded boulderclay of the map must have popped through the coversand.

Usually the symbols for boulderclay are also provided with additional symbols for stoniness. Stony areas unconnected with existing boulderclay (called stonefields above) unfortunately are only rarely indicated on the soil map.

The distances to symbol KX or x (boulderclay at or less than 120cm below the surface) are for the 62 tombs on the available soil map sheets (12E, 16E, 17W, 17E):

for 25	(40%)	0- 50m
for 14	(23%)	75-100m
for 11	(18%)	150-200m
for 9	(15%)	250-300m
for 1	(2%)	550m (D30)
for 1	(2%)	the distance may have been under 300m, but later wind erosion has altered the surroundings (D45).
for 1	(2%)	2500m (D10)

The tombs at a distance of 0-50m from x or KX lie on the brink of boulderclay in the subsoil. For D30 the distance to shallow boulderclay is larger, but in this case symbol '...g' (coarse sand and/or gravel 40-120cm deep), which is 200m away, possibly coincides with a (shallow) stonefield in the sense of the geological map. That large boulders occurred among the gravel is demonstrated by a 'fake hunebed' constructed there during aforestation of the heathlands about 1930 (Klok 1979, ill. on p. 75). D10 lies in a stonefield - in its vicinity the boulderclay is completely eroded.

Again the 300m distance catches the eye. In 59 (95%) of the 61 available cases the distance to the stoney boulderclay was 0-300m. The fact that the necessary capstones and smaller boulders were most probably available within 300m (-400m) according to the maps, explains why there were such good possibilities for hunebed building in the areas where they are situated. The scarcity of hunebeds in the midwest of the Drenthe Plateau must be due to such factors as the greater thickness of coversand there, the lesser erosion of the boulderclay and the possibility that the rare hunebeds were completely removed sooner than in stone-richer regions. In addition, TRB population could have been thinner here because the area was less suitable for agriculture, but this possibility has still to be checked with the soil maps.

10.5 Why the Renfrew game does not work in Drenthe

How are we to understand that Renfrew's Thiessen polygons around megalithic tombs give territory divisions plausible for Rousay and Arran, but much less so for the Drenthe Plateau?

A sensible site location combines as many good properties as possible. The researcher may rediscover what were scarce essential conditions. The scarcest commodities indispensable to the tomb builders' society were critical factors. Their presence determined where the location of tombs, settlement, fields and territories was possible.

On the rocky islands of Arran and Rousay tombs were restricted to the vicinity of the scantily arable soils, while stones for chamber cairn construction were probably plentiful almost everywhere; on the sandy Drenthe Plateau the presence of boulders dictated the place of the tombs, whereas the presence of arable soil was perhaps much less problematic.

Other decisive factors must have been the availability of the manpower needed for the construction of the particular tomb type, and the importance attached by the local society to the building efforts. Each of these may have been different on both islands and the Drenthe Plateau. On the eastern ridges of Drenthe, there is a heavy concentration of hunebeds which suggests that the tombs were built here wherever possible and also as many as possible.

Only in the northwest and the southwest of the Plateau the territory concept seems to work somewhat better, as we have seen above. Probably boulders (and manpower?) were scarcer here. A hint is given by the fact that the three hunebeds D54a-c, which were so exceptionally distant from stonefields in fig. 10.4, belong to the southwestern group. It seems as if the minimally necessary number of tombs were constructed here with much effort. In the northwestern part, for which geological maps are as yet lacking, the situation will not have been much different.

However, it is possible that the regular spacing in both these regions was due to a human territory division which naturally often coincided with the river valleys. But this point needs further reflection because the sides of the regularly spaced river valleys may also have provided the loose boulders for the tombs and the coversand ridges were framed between them.

This line of reasoning deserves clarification of another point. While we saw that perhaps as many hunebeds were built as possible on the eastern ridges of the plateau, there are also a few large stonefield areas without megalithic tombs in fig. 10.4. We will come back to this point in section 9.

10.6 Road alignments of TRB megalithic graves

We mentioned the alignment of many hunebeds on the eastern ridges in section 3. At first sight these alignments can be fully explained by the periglacial topography. In the main they are due to the straight parallel boulderclay/stonefield boundaries visible on the right hand side of fig. 10.4 and the resulting relief and coversand topography (Wieringa 1968, map). Therefore it is unnecessary to postulate alignment along old straight tracks. This does not, however, imply that they were not situated on such straight routes. Several hunebed alignments coincide with the lines much more densely marked by later prehistoric barrows, which certainly were prehistoric routes (Bakker 1976; Jager 1984; Taayke 1985, fig. 10.1). Thus, according to Zipf's 'principle of least effort' these prehistoric roads simply followed the easiest straight course as indicated by the relief (however slight). Thus the hunebed location combined at least (a) in several, not all cases, the side of an important route, (b) a distance not much over 300m from the necessary boulders, (c) as we shall see in the following section, a position on dry sand, usually on a slope or ridge.

Since suitable soils for arable land and settlement sites were available almost everywhere in the drier parts of the hunebed region (sections 7 and 8) the question may be posed whether Thiessen polygons around the tombs are at all helpful to us for detecting territories in TRB society. Were the hunebeds literally genuine central places at all for West Group society? They may quite well have taken a central place in the life of TRB people, without displaying this by their position in the landscape. As markers, they could also be expected, for instance, at the boundaries of fields or at the limits of territories (instead of in the centres of the latter), or, as is even more probable in our case, as monuments or markers along routes.

Recently Kinnes (1975) considered the design of British chambered tomb plans in terms of the linear, dispersed or aggregate arrangement of chamber modules. We think that this tripartition may also be useful for discussing settlement and burial patterns. Kinnes' scheme was reproduced in a monograph on 'The Megalithic Monuments of Western Europe' (Renfrew, ed. 1981, 100). In the same volume Renfrew (1981, fig. 2) discussed a variety of geographical relationships between settlement pattern and the location of burial/ritual facilities - possibly as a way out of a too slavish application by others of his central territory markers concept of the tombs. His seven schematic maps appear to have - or could have - been designed with the help of a combination matrix of agglomerate settlement, dispersed settlement, agglomerate burial/ritual facility (i.e. a major collective tomb) and dispersed burials.

As the alignments of several (but not all) hunebeds on the Drenthe Plateau demonstrate, Renfrew's scheme (1981, fig. 2) should at least be expanded with linear arrangements - as a sort of analogy to the three options of Kinnes.

Location along routes of megaliths, barrows and in Jutland stendynge-graver is a widespread phenomenon in northwestern prehistoric Europe (Bakker 1979) which is essentially incompatible with the concept of megaliths as central territorial markers.

In the same volume, Kaelas (1981) gives a map compiled by Hyenstrand of over 250 megalithic passage graves in an area of some 1350 sq km in Fallbygden in northern Västergotland, Sweden (fig. 10.5). Their density (on average one tomb per 5.3 sq km) is about seven times that in the area of the Drenthe Plateau in fig. 10.1 (on average one tomb per 36.5 sq km).[6] The tombs show partly clustered, partly dispersed distributions, and alignment along roads (which take account of the numerous swamps and the relief) is also striking. The area was subdivided by Hyenstrand by straight lines into territories with 7-25 tombs each, based on the assumption that an immediate area of 2 - 2.5km radius from each passage grave group was the supply area for food production.

Hyenstrand's territories seem to me as artificial as Thiessen polygons would have been in this case. The impression one gets when the territory divisions are omitted from Hyenstrand's map is one of diffuse road alignments which clot together in many places where the number of tombs becomes large and they begin to cluster. A map of over 950 barrows in the Veluwe (central Netherlands) has the same effect (Klok 1982).

A map by Schlicht (1968, fig. 10.6) of the megalithic graves in the German Hümmling directly east of fig. 10.1 displays a much more distinct set of route alignments than those of Fallbygden and the Drenthe Plateau. The known megaliths are here hardly clustered and almost all seem to lie on ancient roads which take the straightest course and avoid crossing the peaty river valleys as much as possible. Where they crossed valleys, the prehistoric routes are often still in use to the present day and at the ford at Spahn-Werpeloh there are late medieval defensive earthworks (Landwehren, Schlicht, pers. inf.). Application of Thiessen polygons to this aligned distribution pattern gives no convincing results at all.

Instead of relying too much on Thiessen polygon territories in our regions, research could better concentrate on the mapping of the distribution of individual or very local features in the pottery and thus derive, eventually, insight into the interaction spheres at the local level, i.e. of the communal areas (or 'territories').

10.7 Coversand types

The 1:50,000 soil map reveals a great many details in the drier coversand surfaces which were not covered by peaty or humic-sandy gley soils. Code names run like Y23-VII, Hd21-VII, Hn21-VI, Hn23x-V, KX23-III, but there are many more combinations of the following three sorts of features.[7]

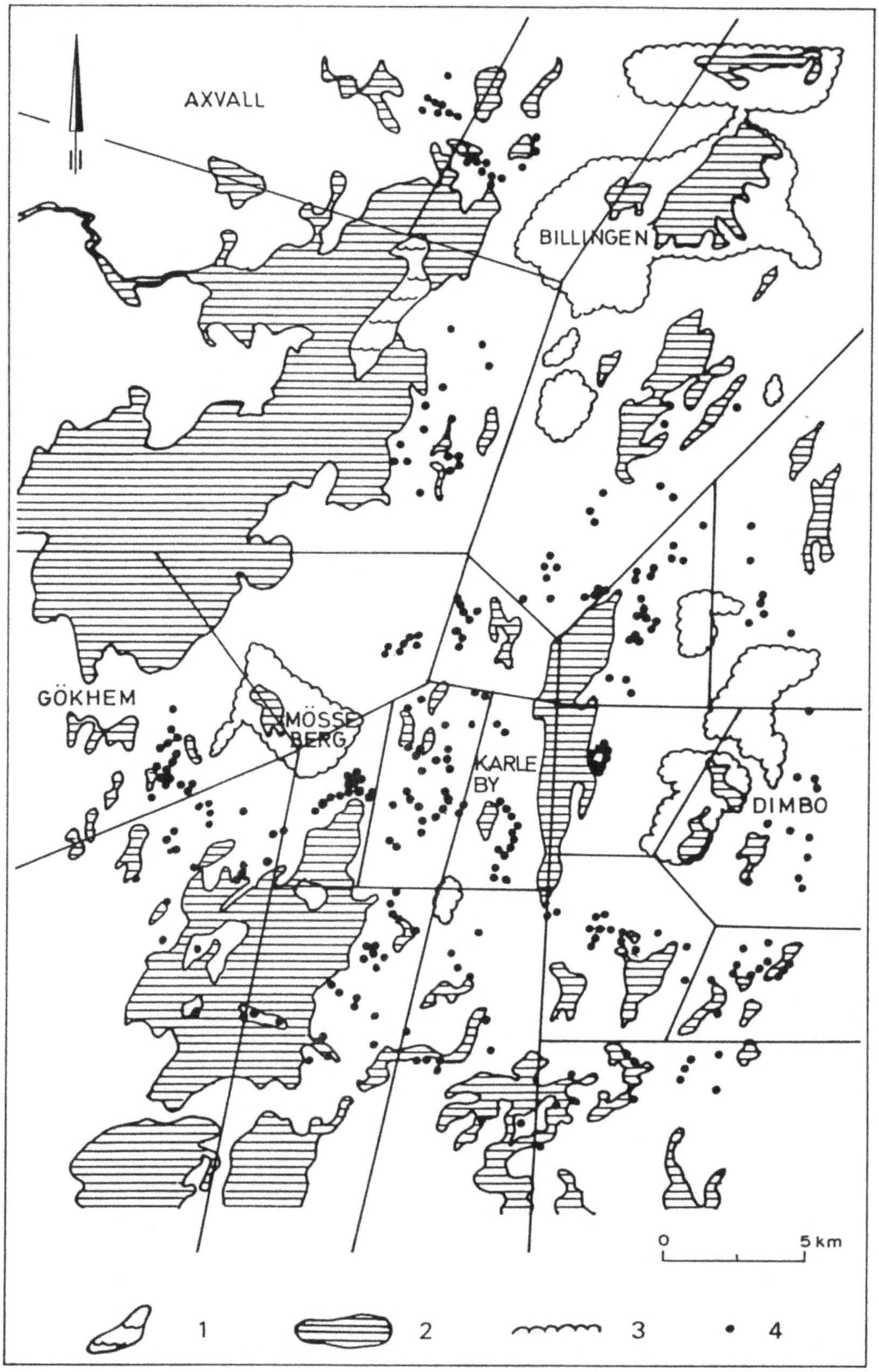

Fig. 10.5 Hyenstrand's territorial division of the Fallbygden passage grave area, Sweden. When the boundary lines are omitted, several road alignments emerge.
1 = lake 2 = marshland
3 = plateau 4 = passage grave
Source: Kaelas 1981, fig. 10.

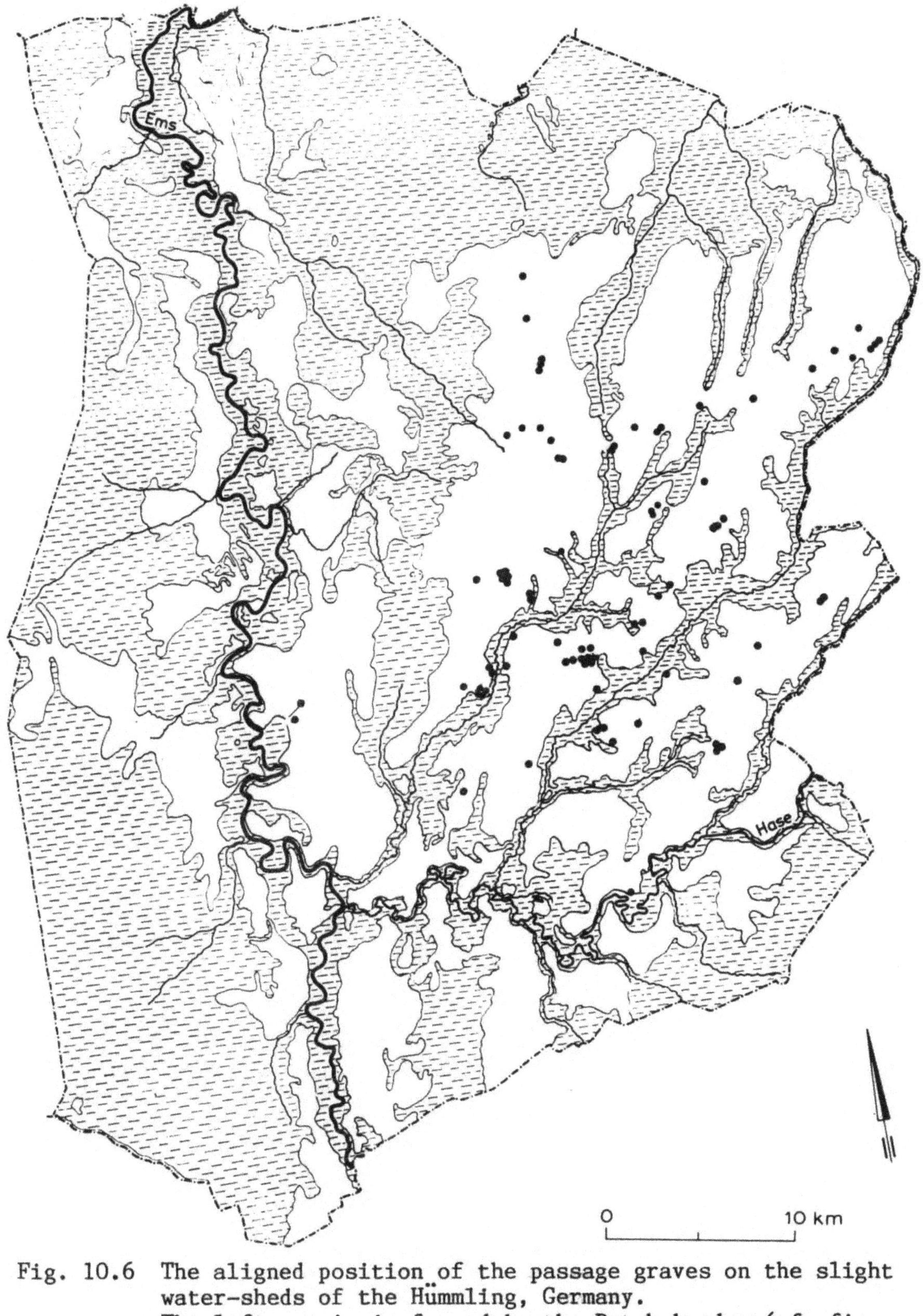

Fig. 10.6 The aligned position of the passage graves on the slight water-sheds of the Hümmling, Germany.
The left margin is formed by the Dutch border (cf. fig. 10.1) and the tombs are 30 to 50km east of the right margin of fig. 10.2. The geomorphology of this landscape is very similar to that of the Drenthe Plateau. The peaty deposits in periglacial valleys and bogs are shaded.
Source: Schlicht 1968.

(a) soil type sensu stricto. The most important are:

Y = holtpodzol, 'typical (non-humic) brown podzolic earth' (Avery 1973)
Hd = haarpodzol, 'humuspodzol'
Hn = veldpodzol, 'typical (humic) gley soil'
pZg or pZn = beekeerd, broekeerd, gooreerd, 'typical humic-sandy gley soil'
KX = boulder clay or 'potclay' less than 40cm below surface
x (added to the code name) = boulderclay at 40-120cm below surface

(b) texture of the upper layer of parent material:

30 = coarse
23 = loamy fine sand (more than 17.5% loam)
21 = fine sand with little or no loam (less than 17.5%)
('loam' is the fraction smaller than 50μm)

(c) main ground water type as determined by the Mean Highest Groundwater and the Mean Lowest Groundwater levels below the surface:

VII	=	MHG	>80cm	MLG	>120cm
VI	=	MHG	40-80cm	MLG	>120cm
V	=	MHG	<40 cm	MLG	>120cm
IV	=	MHG	>40 cm	MLG	80-120cm
III	=	MHG	<40 cm	MLG	80-120cm
II	=	MHG	(<40 cm)	MLG	50- 80cm
I	=	MHG	(<20 cm)	MLG	<50cm

Fig. 10.7d compiles the natural present distribution of soil properties in the southeasternmost part of the Hondsrug (the eastern ridge in fig. 10.1). It is based on a systematic sample taken at the crossing of the kilometre co-ordinates. Peaty and beekeerd/broekeerd/gooreerd soils were left out of consideration. Though 21, VII and soils without x are more common than the other properties, Hd is less common than Hn.

Fig. 10.7a compiles the soil properties on the 59 hunebed sites on the available soil map sheets 12E, 17W and 17E.[8] The preference for Hd, 21 and VII (without x) is striking. The same is true for the 15 flat grave sites and the 40 settlements or findspots of stray sherds considered as such. Two other researchers have confirmed this finding for the hunebed sites by an independent approach (Hegener 1983; Huiskes 1985).

How are we to understand this remarkable preference of the hunebed builders? Were they as good soil scientists as the pedologists of the Wageningen Soil Survey? Perhaps as expert farmers they were, but other factors directly related to the named soil properties may also have been involved. The type of vegetation could have been closely bound to the details of the present soil map, for instance. I once tried to explain the apparent TRB preference for Hd21-VII in this way (Bakker 1980). The local coversand genesis and topography seem to present a better solution, however.

First, it should be noted that the present soil types Y, Hd and Hn may largely have developed only after TRB times on the higher parts of the plateau where the hunebeds are found. The ground water type may also have changed considerably since TRB periods.

Since TRB times, the sea level has risen 3.5m, which caused the drowning of the lower parts of the Plateau, a rise of the ground water table on the remainder and an expansion of peat in the valleys. As the plateau tilts upwards from the northwest, that corner was affected most. Conversely, the highest remnant, in the south-east corner, was least altered and its present condition gives us the most realistic idea of the relative soil moisture in TRB times. Figs. 10.7a-c show that sites on sheets 12E and 17W were already displaying a slight drowning effect during the TRB period.[9] The avoidance of VI-I and of the connected soil type Hn on the southeastern sheet 17E should therefore not be considered as an 'iron law' for TRB site location at the lower parts of the plateau on the other map sheets; of course Hd, 21 and VII must also have been highly preferred there, but many soils which would have had these codes on sheet 17E may have become wet afterwards and now carry codes like Hn and VI-I, or have been covered by peat or clay. Also actual Hd21-VII coversand ridges have become so narrowed by the drowning that they are omitted on the soil map. The geomorphological map may be useful, especially there.

The Toelichtingen (explanation books) for the different 1:50,000 soil and geological maps sheets (Stolp 1977; Booij and van Oosten 1978; Ter Wee 1979) describe the occurrence of the coversands in Drenthe. Elsewhere in the Netherlands the usual stratigraphical series from bottom to top is Old Coversand - Young Coversand I (with Allerød soil horizon at the top) - Young Coversand II. The latter, which is the most common on the Drenthe Plateau, normally contains no loam. Young Coversand I usually contains some loam, but the Drenthian Young Coversands, rich in loam, would find their counterpart elsewhere in the Old Coversands - which, however, hardly occur on the Drenthe Plateau. Booij and van Oosten (1978, 38) explain the loaminess of the Young Coversands by assuming "that shallow keizand and probably also some boulderclay were mixed with the Young Coversand by kryoturbation and other frost action". Keizand is the sandy erosion residual in stonefields or on top of the boulderclay.

This is an important observation because it connects the loamy coversands (23) directly to a substrate of boulderclay. It follows that they cannot be found on top of completely eroded boulderclay ('stonefields') or in stream valleys (fig. 10.4). This is quite different from what is the case with loamy Old Coversands (outside the

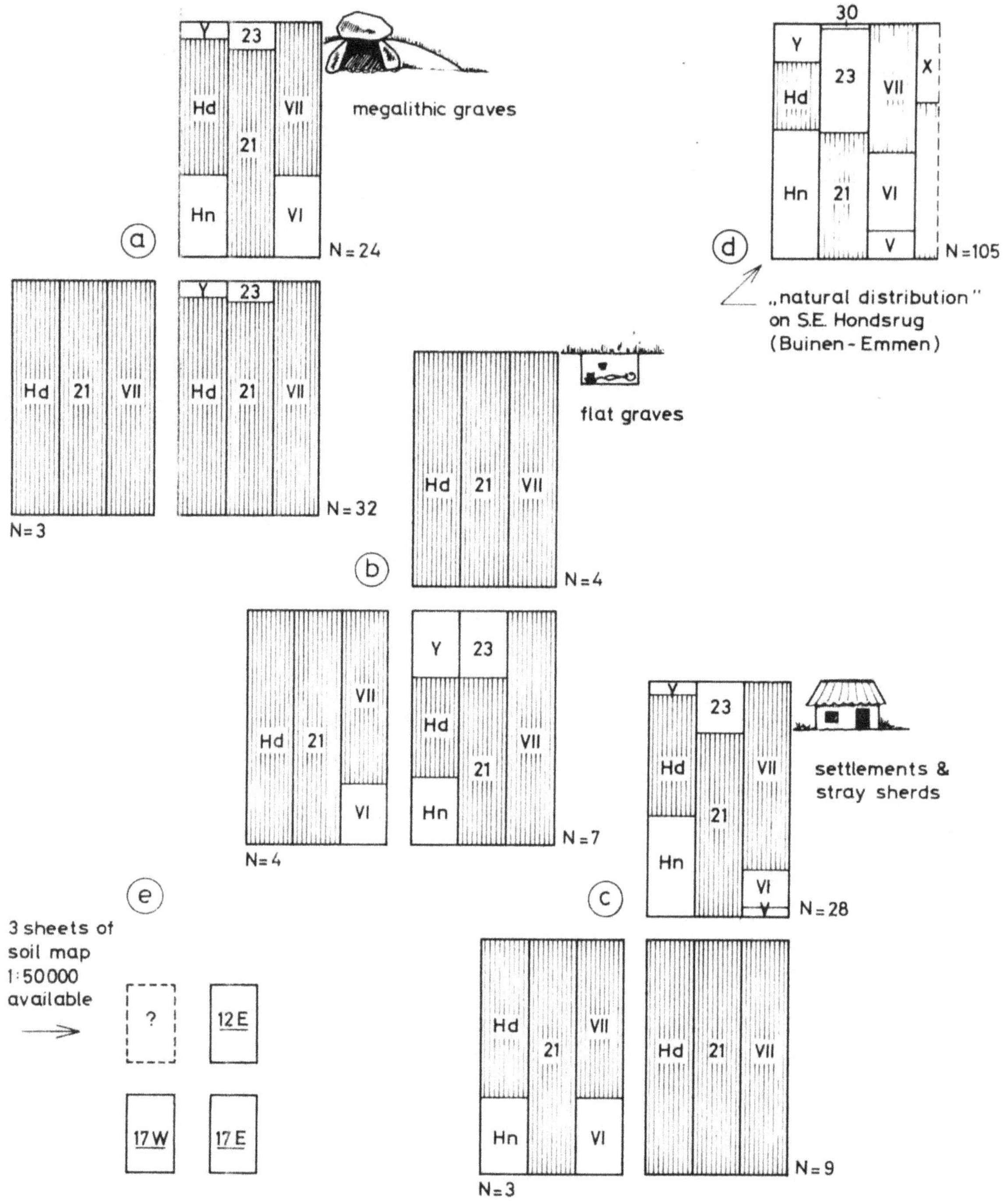

Fig. 10.7 Bar graphs of site characteristics of (a) hunebeds, (b) flat graves, (c) settlements or stray sherds on the Drenthe Plateau, according to the sheets 12E, 17W and 17E of the 1:50,000 soil map (e). The present 'natural distribution' of these features on the driest southeastern sandy 'island' of the Plateau is rendered by (d). The number of observations is shown by 'N'.
Source: Bakker 1980, 1982, with small corrections.
Drawing: IPP.

eastern Drenthe Plateau) which were as such deposited by the wind and the loaminess of which was not caused by secondary kryoturbational mixing with its substrate, as in eastern Drenthe.

The usual coversand thickness on the eastern ridges is 60-150cm. Especially the 40-120cm thick coversand layers on top of the boulder-clay are loamy (code 23x).

The narrow ribbon coversand dune ridges on which so many hunebeds lie are much thicker than 120cm and usually contain no loam. Some of these ridges, however, are loamy (for instance the ridge on which tomb D31a lies, according to Wieringa 1968).

Whether a boulderclay substrate 40-120cm deep (23x) causes stagnating groundwater and a moist sand cover (of type VI, V or less) depends on the slope of the surface of the boulderclay. While stagnating groundwater is usual on top of till plateaus or in slight depressions, steeper sloping boulderclay surfaces may carry well drained sands (VII).

The site location of most hunebeds on soils with code Hd21-VII (without x) may - apart from the necessity of nearby boulders, a dry sandy subsoil and an unknown relation to the settled and reclaimed area - be due to a slight preference for a certain morphology of the landscape. Wieringa (1968, 155) - whose pioneer research of the relations between hunebeds, subsoil and landscape has stimulated the present study - describes the position of 25 tombs on the southeastern Hondsrug on map sheet 17E as follows (fig. 10.8):

(a)	1	(4 %)	on a coversand ridge on top of a boulderclay plateau
(b)	10	(40%)	on a coversand ridge on the flank of a hill
(c)	3	(12%)	on a coversand ridge at the foot of a slope towards an erosion valley or stream valley
(d)	5	(20%)	on a slope of eroded boulderclay with a thin layer of fluvioglacial material (a stonefield) on top of extremely fine proglacial sand
(c)	6	(24%)	on a layer of coversand on a flank of a hill.[10]

Thus all 25 tombs 'would have had a view' if there were no trees in the way, lying on a slope, the top of a ridge, or both.

All lay on dry sand. 20 (80%) lay on coversand and five (20%) on a similarly dry and very fine, but much older sand (d). The concept-soil map shows that in 24 (96%) cases the soil was Hd21-VII (including group d). The only exception (D31a) had Hd23-VII. It belongs to group (b), which implies that there are also coversand ridges which are loamy, as mentioned above.

None of Wieringa's examples lies on boulderclay within 40-120cm below the surface. This is almost always the case in Drenthe.[11]

As we have seen above, the hunebeds were situated on Hd21-VII, generally somewhat elevated above their near surroundings. The limited extent of the clearings around the megaliths must, however, have

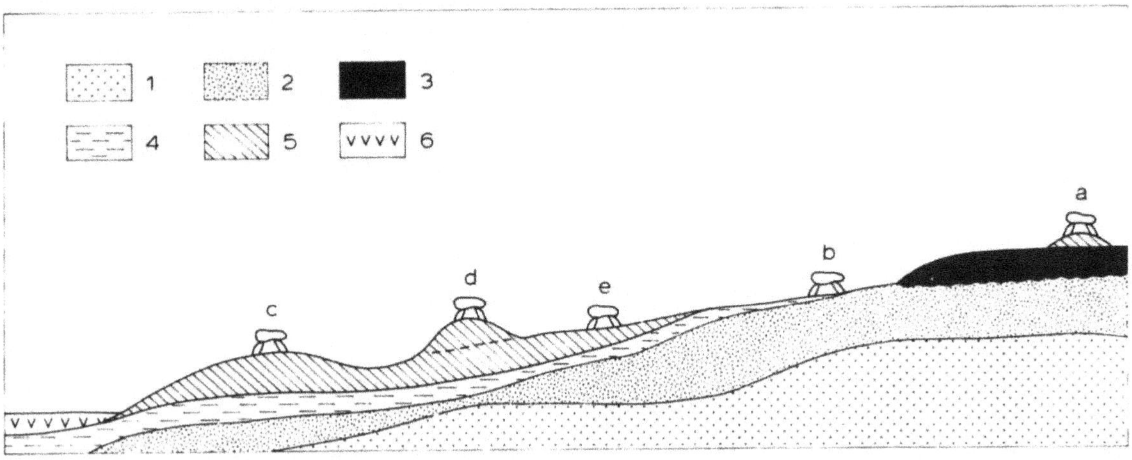

Fig. 10.8 The different positions of Drenthian hunebeds.
1 = pre-glacial coarse sand and gravel
2 = pro-glacial fine sand
3 = boulderclay
4 = fluvioglacial residual of eroded boulderclay etc.
5 = coversand
6 = peat
Source: Wieringa 1968 (redrawn at IPP).

precluded better visibility from afar as the major determinant. The main purpose was, more probably, better drainage of the monument, cf. O'Kelly's description (1982, esp. 100–101) of all the efforts made by the builders of the Newgrange chamber tomb to keep the inside dry. Well preserved TRB megalithic graves in the North group have loam and stone packings which lead the rainwater from the capstones aside. By such measures the chamber is effectively kept dry. In the West group ruins of stone and loam mantles may be interpreted likewise. The gravels of burnt flint or burnt granite covering the chamber floor pavement must also have provided drainage. Modern cemeteries, of course, are also laid out on dry ground.

10.8 Soil, vegetation and the influence of TRB man on the landscape

Soil samples from old ground surfaces underneath 12 hunebed barrows have been analysed for their pollen content (for reference to the original data, see Casparie and Groenman-van Waateringe 1980, N1–N12). Sod samples from two other hunebed barrows (N13–N14, Diever-D52 and Wapse-D52a) were apparently of a younger date because Fagus pollen (which was completely absent from the other samples) occurred in them, and will not be used here. In Table 10.1 the rough pollen data have been corrected according to Iversen (1964; cf. also Andersen 1970, 1980 and 1985) for pollen production and according to Havinga (1984, Table 10.2, the data given under the heading Podzol) for differential corrosion. Herbaceous pollen types were left out of consideration, except for Gramineae and Ericaceae, because no correction factors for pollen production and differential preservation are available.

In fig. 10.9a the pollen data are grouped according to (1) trees belonging to the climax wood vegetation, (2) trees from a damp habitat and (3) pioneer trees on abandoned clearings (see, however, below concerning the birch). Furthermore, data from grasses and heather are given which were the main food plants for domestic animals and formed the main herbaceous vegetation cover (cf. Groenman-van Waateringe 1986).

While the hunebeds analysed on the areas with soil maps are generally situated on soil type Hd21-VII, the differences in pollen content of the old ground surface samples must have been caused by human behaviour and the vegetational history of the site and its immediate surroundings. Therefore the soil types present within a radius of 500m around each megalith were analysed, as given by the concept sheets 17E and 12E of the 1:50,000 soil map (fig. 10.10). No such map is yet available for hunebed D1.

Correlation of pollen data with percentages of different soil properties (Hd vs. Hn, H vs. Y, 21 vs. 23, VII vs. VI-V, etc.) seems to lead nowhere (fig. 10.9b).

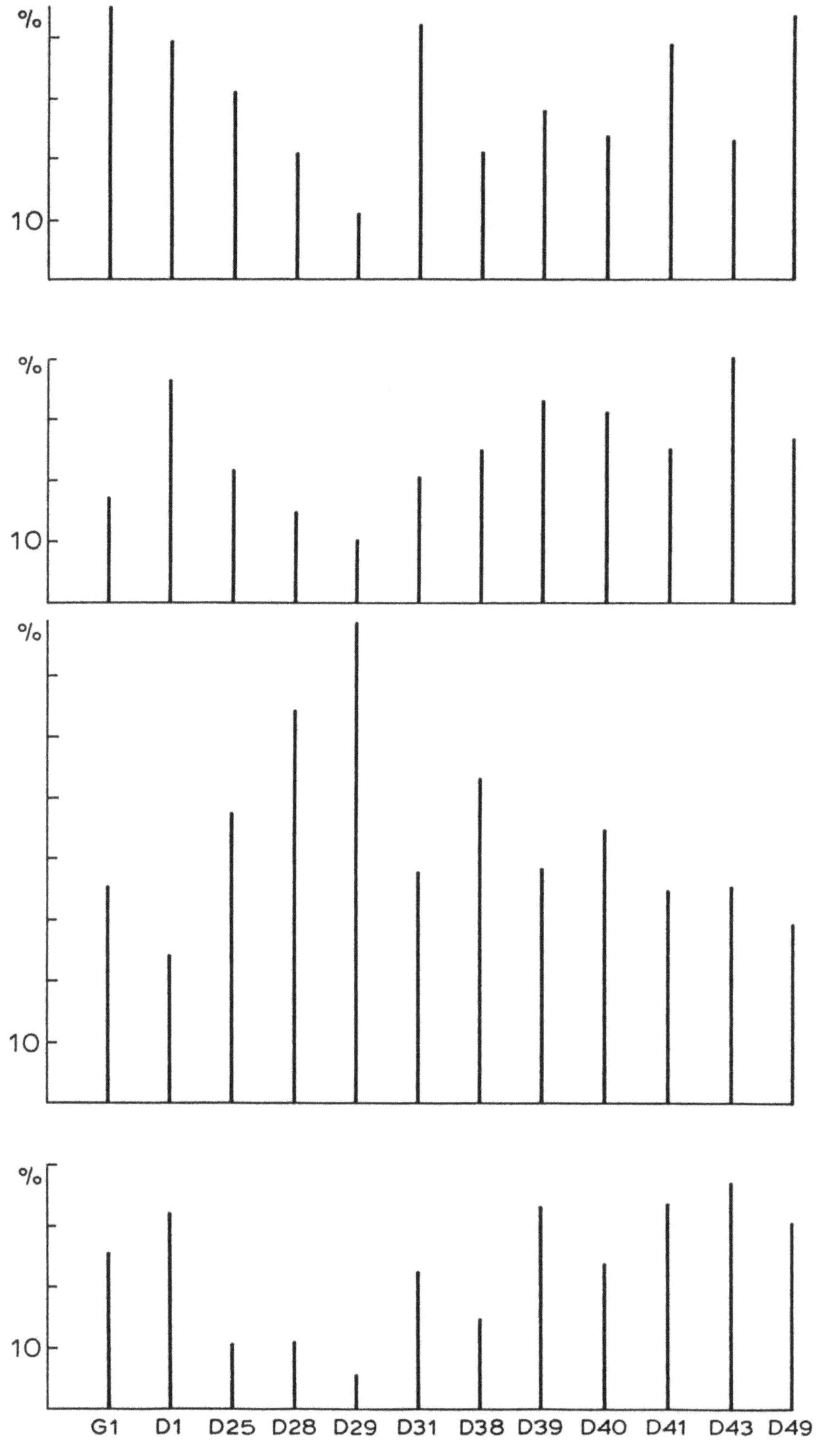

Fig. 10.9a From top to bottom: corrected data for (1) Quercus, Ulmus, Tilia and Fraxinus; (2) Alnus and Salix; (3) Betula and Corylus; (4) Ericaceae and Gramineae. Drawing: IPP.

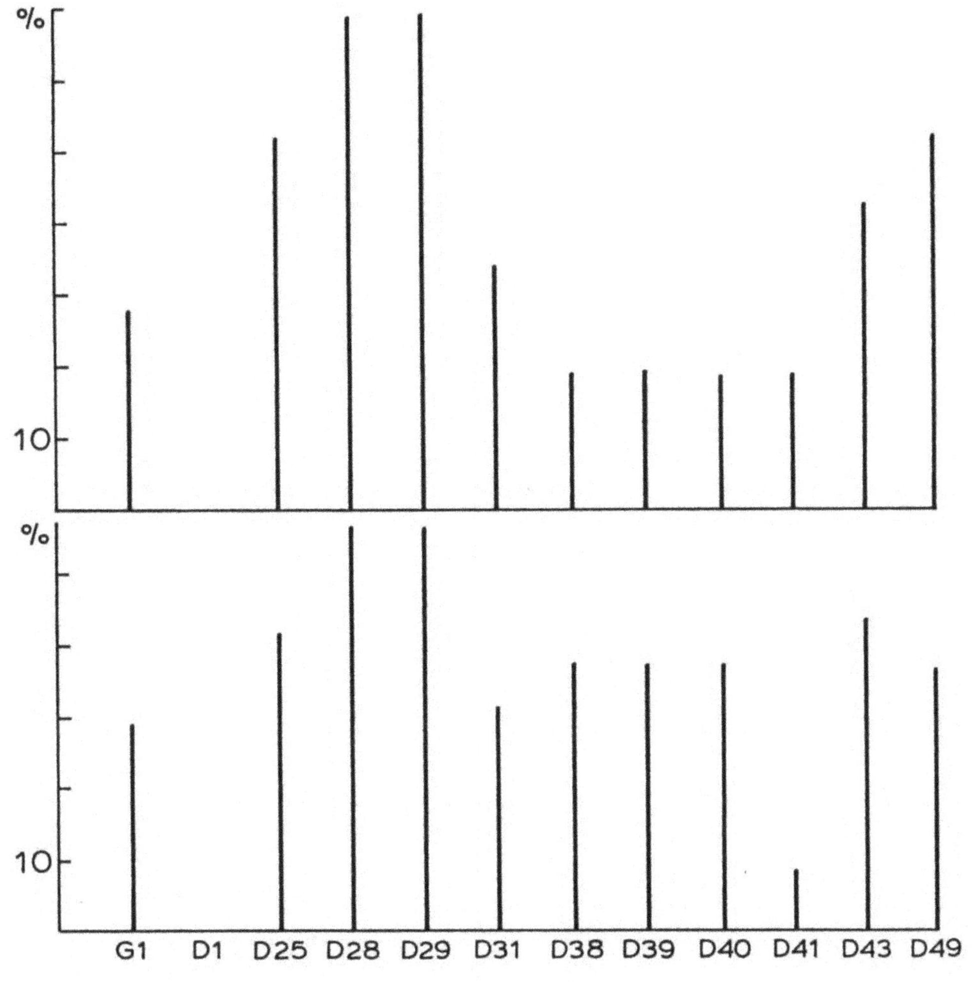

Fig. 10.9b Upper: percentages of soil types Hn plus ground water levels VI and V.
Lower: percentages of soils with 23 texture.
Drawing: IPP.

The only apparent correlation is by the percentages of Tilia and Fraxinus with the distance from the code 23 soils in SW and NW direction, which are the main wind directions in the Netherlands: **Hunebed samples with high lime and ash values lie further away from code 23 soils than those with low values** (Table 10.1, figs. 10.11-12). Code 23 indicates the loamy fine coversands, whereas 21 indicates fine sands without loam or poor in loam, as we have seen above. Neither contains any stones.

Lime and ash prefer somewhat richer soils, but well-drained and with a loose structure. A podzolized soil (Hd) is unsuitable for Tilia, but we may assume that leaching of the soil had not yet occurred on a large scale on these dry sandy soils, so that the present difference between code Y, for holtpodzols (or 'typical non-humic brown podzolic earths'), and Hd, for haarpodzols (or 'humuspodzols'), did not yet exist at the time. We may expect Tilia to have occupied the place in the climax forest taken later by Fagus. The later competition between man as cultivator and Fagus, both preferring the same richer soil types, has been well demonstrated (cf. Burrichter 1976, Aaby 1986). Similarly the Neolithic farmer must have looked especially for woods where Tilia abounded, knowing by experience that the richer/ loamier, loose and well drained soils under such woods offered the best possibilities for his agricultural activities.

Low Tilia values may thus be due (1) to a rare occurrence of lime because of podzolization (Hd), and (2) to an intensive use mady by the farmers of the loamier coversands (23) where lime was the main constituent of the wood. Since podzolization had not yet generally occurred on the drier sandy soils, the second cause must have been the prime determinant. This means that TRB people mainly attacked the woods on 23 soils to use them for their fields.

Tilia is insect-pollinated which implies that its pollen in the spectra must derive from the close neighbourhood of the tombs. The observed - and at first sight paradoxical - correlation between high Tilia and Fraxinus values in Table 10.1 and a distance of over 100m to the loamy code 23 (arable) soils in SW and NW direction in Table 10.2 cannot therefore be due to those distant 23/lime woods which had partly been cut by man. Thus the Tilia pollen must derive from the scarcer lime trees near the hunebeds, growing on the Hd21-VII soils. Where the loamy code 23 soils occurred closer than 100m to the hunebeds, Table 10.2 shows low Tilia values which can only mean that the original wood with its dominance of lime trees was substantially cleared by man to bring the loamy soil into cultivation. Felling of lime trees more than 100m away would not be reflected in a local pollen spectrum, whereas a similar amount of felling much closer would influence the percentages of different tree pollen considerably.

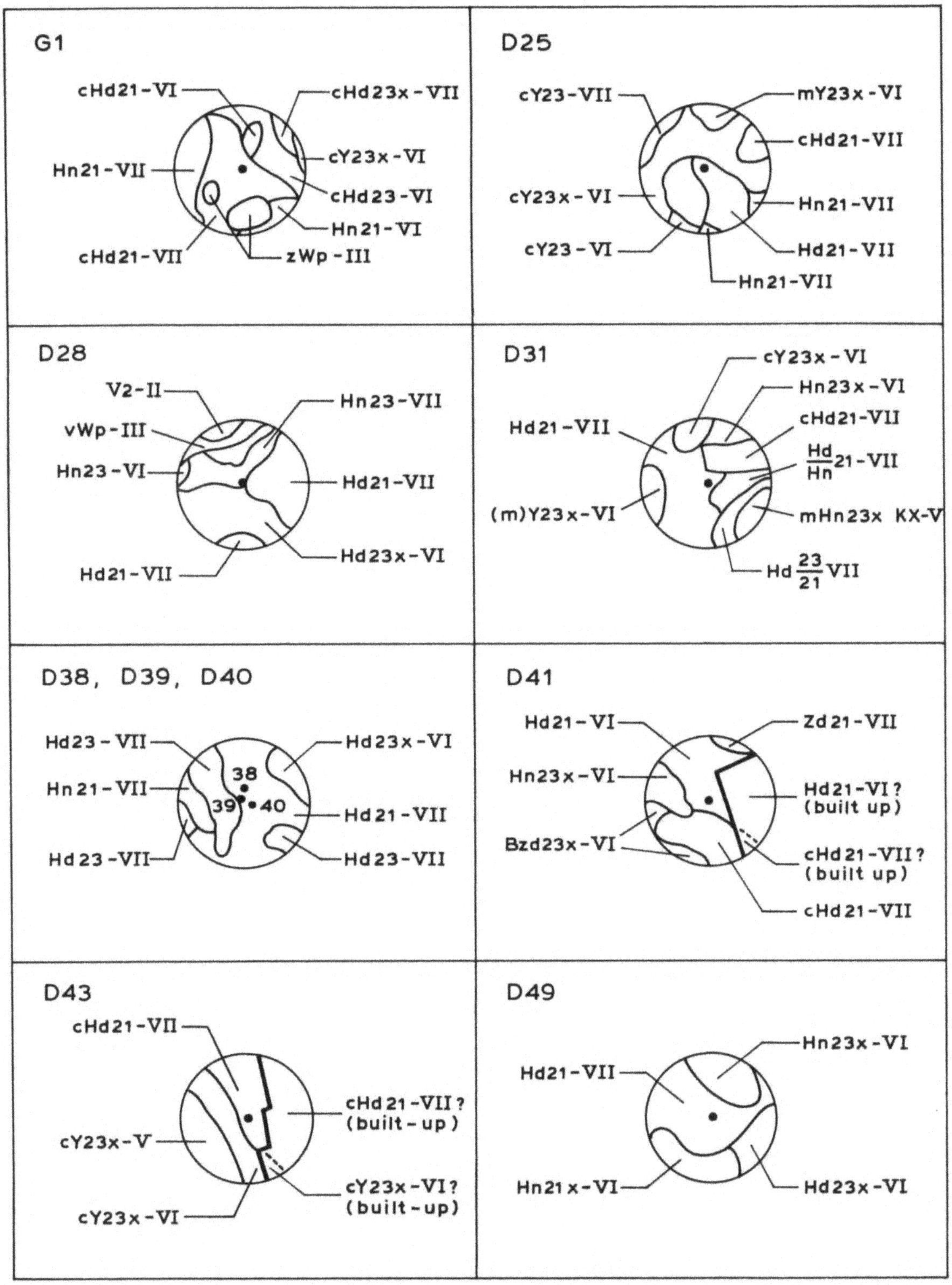

Fig. 10.10 The soil types within a 500m radius around 10 hunebeds according to the concept maps of sheets 12E and 17E of the soil map. Usually the concept is more detailed than the published map, though also later corrections may have been used for the latter. Drawing: IPP.

Table 10.1 Corrected pollen percentages

		G1	D1	D25	D28	D29	D31	D38	D39	D40	D41	D43	D49
Alnus	0.4	16.3	33.3	15.9	14.4	10.6	18.6	25.0	32.9	31.1	24.7	34.3	21.6
Salix	3.4	1.8	3.1	5.3	-	-	1.5	-	-	-	-	5.9	5.0
Corylus	0.5	11.7	16.9	17.1	9.2	3.1	18.3	15.4	24.0	19.8	15.7	22.3	13.5
Betula	0.7	23.7	7.0	30.5	55.4	75.2	19.0	37.8	14.6	25.0	19.5	13.0	16.0
Quercus	1.5	13.1	6.9	10.6	15.0	7.0	10.5	10.4	24.9	11.3	17.4	15.7	14.7
Ulmus	1.2	0.1	0.6	0.5	-	-	0.3	-	-	-	0.3	2.1	2.1
Tilia	2.6	29.4	29.7	19.4	5.5	4.1	18.3	11.0	3.0	12.4	18.6	6.0	21.7
Fraxinus	5.4	2.8	2.5	-	-	-	4.8	-	-	-	2.7	-	5.3
Pinus	0.3	0.03	0.1	0.7	0.5	-	0.9	0.4	0.6	0.5	1.1	0.7	0.3
Ericaceae	0.25	12.4	7.1	6.7	8.7	3.6	14.8	10.9	15.8	19.1	26.2	18.8	28.3
Gramineae	0.5	13.4	24.9	4.5	2.6	2.3	8.4	3.9	17.2	5.0	7.5	18.6	2.0

Correction factors are given after the name of the pollen type

Betula is included in the pollen sum

Table 10.2. Distance to soils with a 23 texture

	<u>G1</u>	<u>D25</u>	D28	D29	<u>D31</u>	D38	D39	D40	<u>D41</u>	D43	<u>D49</u>
SW	-	400	0	0	300	100	50	50	200	75	-
NW	-	200	0	0	300	-	0	0	-	75	-
NE	100	50	0	0	450	200	350	350	-	-	150
SE	-	-	0	0	75	400	300	300	-	75	250

- = further away than 500m

underlined mound numbers have high Tilia values

Harsema (1982, 151-153) supposed that the inhabitants of the well known TRB settlement of Anloo used nearby Y23x-VI soils for their arable land. That the fields were indeed situated on the code 23 soils is shown by the correlation between the low Tilia values and the added percentages of Cerealia plus Plantago lanceolata >1 on the one hand, and values for cereals and plantain under 1% with high Tilia on the other (fig. 10.11). Exceptions are D1, for which the soil map is not yet ready, and D29, of which the pollen sum is too low to be trustworthy in this respect. The difference in the arboreal pollen (AP) percentages (fig. 10.13) may also point to more open vegetation, i.e. fields nearby, correlated with low Tilia values.

The forest around the tombs D28-29, which lie next to each other, was dominated by Betula, not so much as a pioneer tree following the abandonment of a clearing, but as a result of soil conditions: a position at the edge of soil types Hd21-VII, Hn23-VII and Hd23-VI favours rather damp birch wood. The high values for Dryopteris type and Sphagnum spores point in the same direction. In the forest around D39, Quercus and Corylus must have been the dominant trees. D43 has the highest values for alder: its situation directly next to moist Y23x-VI and down the wind of the wet Y23x-V soil of the Emmeres boulderclay ridge may explain this.

10.9 Final remarks

The interpretation given in section 8 solves the existing discrepancy between the reasonable assumption that loamy coversands were preferred for TRB arable soils and the fact that pollen spectra from old ground surfaces below hunebed barrows often showed cereal pollen. Thusfar the latter observation was understood as demonstrating that hunebeds were built in fallow arable fields, which meant that there would have existed a second type of TRB arable land, made on dry and loamless Hd21-VII sands.

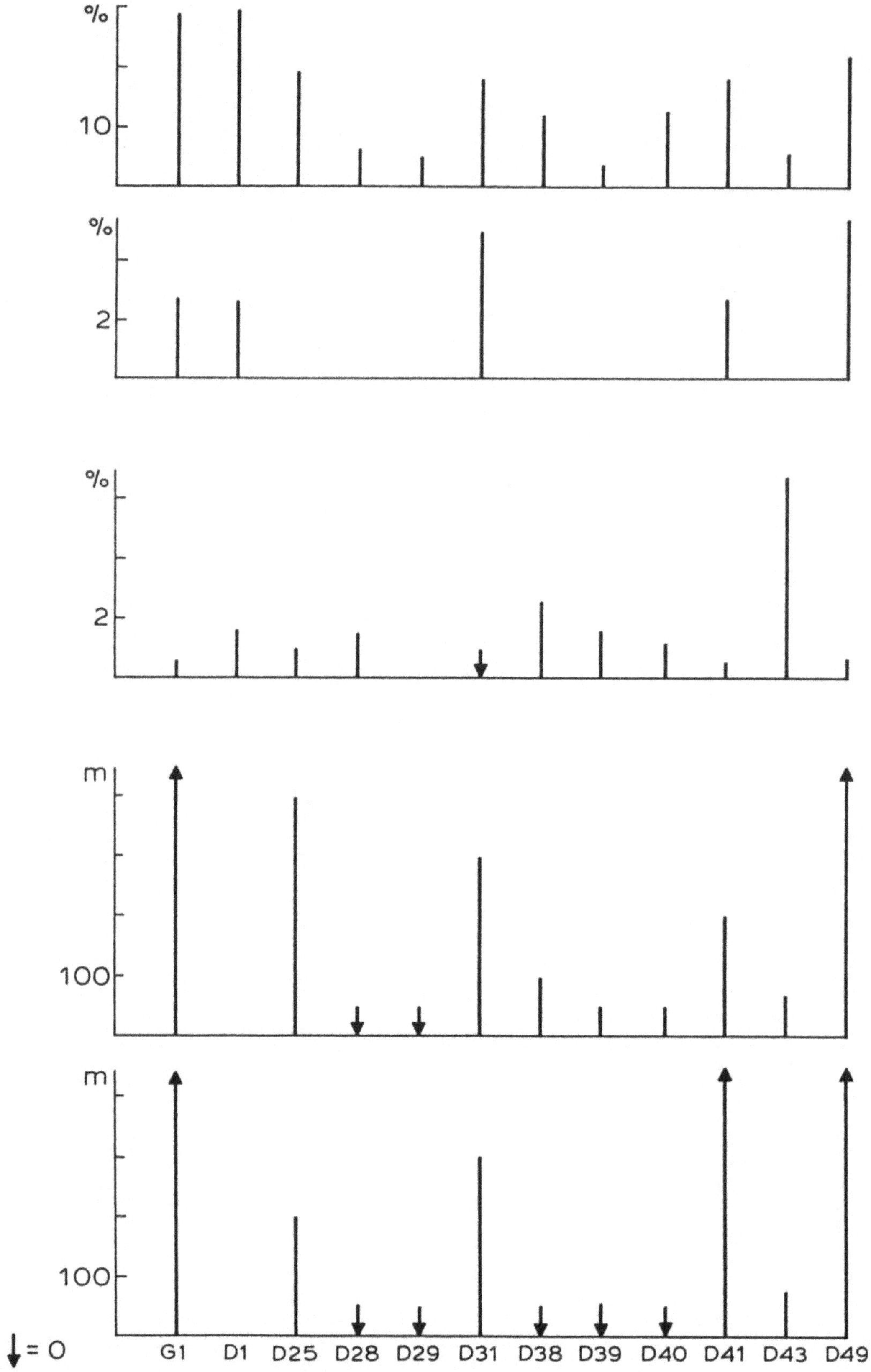

Fig. 10.11 From top to bottom: corrected data for (1) Tilia; (2) Fraxinus; (3) Cerealia plus Plantago lanceolata (data are percentages of the new Arboreal Pollen sum); (4) distance in m to 23 soils in SW direction; (5) idem in NW direction. Drawing: IPP.

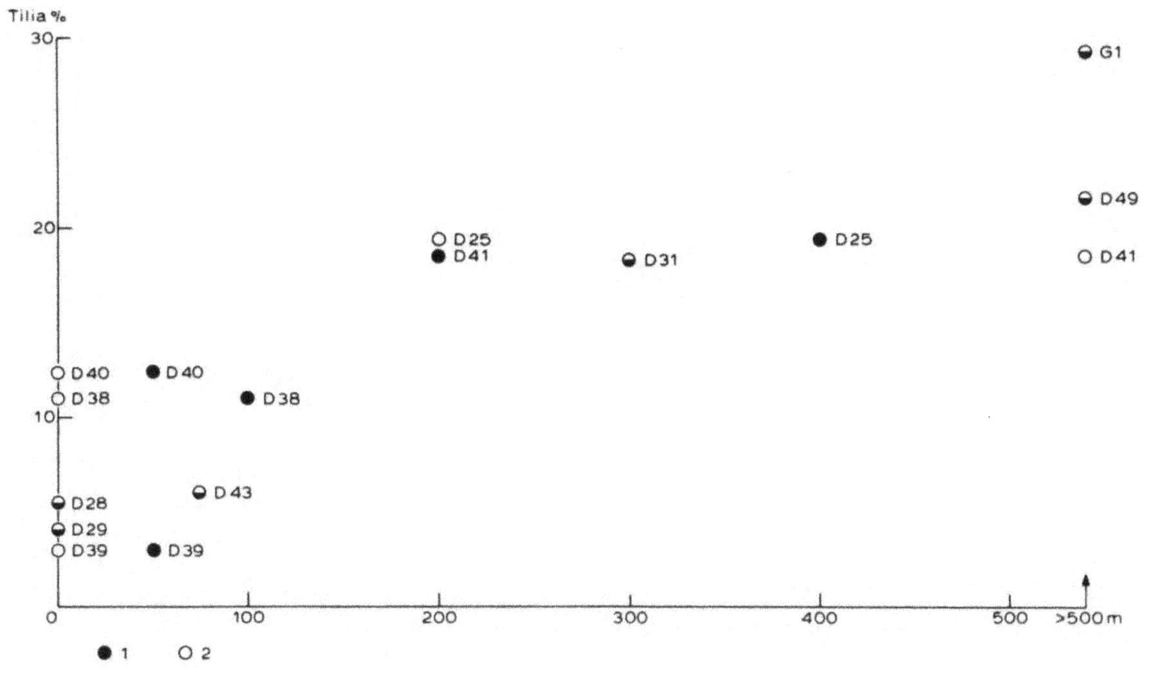

Fig. 10.12 Distance in m to soils in a SW and a NW direction in relation to Tilia percentages (1 = SW direction, 2 = NW direction). Drawing: IPP.

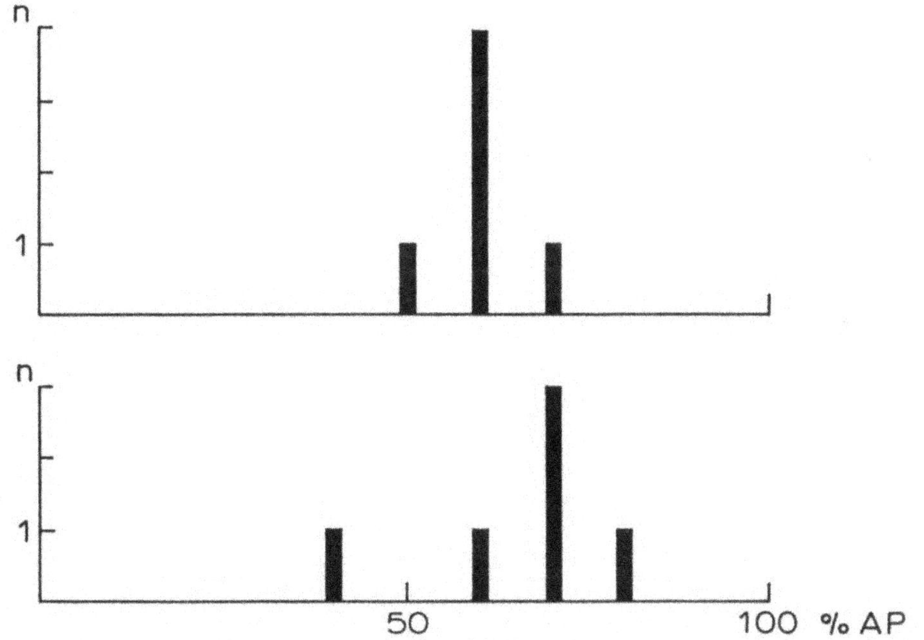

Fig. 10.13 Arboreal pollen percentages: upper for the old ground surfaces of the hunebed barrows with low Tilia percentages, lower for high Tilia percentages. Drawing: IPP.

This rather complicated and incomprehensible model has now been rendered obsolete. Hunebed construction preferably took place on the present soil type Hd21-VII whereas arable land was preferably cleared on such present soil types as Y or Hd23-VI or VII. The cereal pollen from the distant fields was blown towards the hunebed building site clearings.

How we may picture the hunebed sites in the forest is another matter. The ceremonies which took place within and in front of the tombs over the centuries presuppose a larger parvis than just the area where pottery was ceremonially deposited, reaching less than 25m south of the entrance.

We do not know if all members of the local society took an active part in these ceremonies, if there were onlookers from a profane distance, if visiting parties camped in the clearing, and so on.

Were the tombs still enclosed from behind by the imposing primeval forest trees and were they visible only from the trackways along which they were often situated?

In sections 2-7 nothing was found to show that hunebeds actually lay in the centre of the territories of local communities (as Renfrew suggested for other megalithic landscapes). Alignment along traffic routes seemed, however, often probable.

In addition, statistical relations between hunebed sites and geological, pedological and geomorphological features of the landscape were investigated. Such relations, however convincing statistically, may sometimes make no sense at all. The black woodpecker, for instance, has a distribution in Holland north of the Rhine that is exactly congruent to the TRB area. The link between both distribution patterns is the pleistocene sandy area once settled by TRB man and attractive to black woodpeckers because it now abounds in pine woods and old beech trees full of holes suitable for their nesting (Bakker 1982).

Less obviously, indirect relations may also cloud our perception of the reasons for the choice of TRB sites. Higher parts of Drenthe were especially used for hunebed building, but as we have seen this was greatly determined by the greater erosion of the boulderclay in such parts in periglacial times. The profusion of boulders made these regions attractive to hunebed builders, as most probably did the diversity of these landscapes with their patchy boulderclay substrate. The presence of dry, elevated and loamless sandy soils for hunebed sites, not too far away from stonefields or arable fields on loamy sands must have formed a main attraction.

Application of statistical relations from the region where they were derived, to another, (for instance to explain the absence of hunebeds) brings the correctness and the predictive value of those direct relations to a test. Whether the northwestern tip of the Drenthe Plateau (fig. 10.1) was indeed always without hunebeds, and if it had a less dense TRB settlement than the Hondsrug area is at present uncertain. Where, or even whether, the supposed attractions for hunebed building and agriculture were present may only be answered after the appearance of the relevant geomaps.[9] Yet the possibility remains that an area with all the named attractions was still unattractive due to one or more other, unknown factors.

That such answers cannot yet be given easily even in areas for which the relevant maps are published, is shown by a final exercise. Fig. 10.4 shows that there are some areas on the southeastern plateau where the boulderclay was completely eroded, but where no hunebed sites are known. They are:

(1) north of hunebed D49 the long strip of the Ellertsveld, and the adjoining Wezuperveld and Westdorperveld (grid references 245/538-550).

(2) the Noordbargerbos (245/532).

(3) the Eeserveen-Odoornerveld (250/541-548).

In area (1) VII is rare and there are mostly wet Hn sands. Hd21-VII soils occupy relatively small surfaces. Code 23 soils are, however, also present and it cannot be said that situations like those around tomb D49 are completely absent. Half of area (2) is occupied by Hd21-VII soils, the other half by Hn21-VII/VI/V, but 23 soils are absent. In area (3) Hn21 soils prevail and there is a long undulating coversand ridge with Hd21-VII. Code 23 soils are however rare.

Fig. 10.4 suggests that what these three areas have in common is that almost everywhere the boulderclay is completely eroded. Although some remnants occur in areas (1) and (3), their greater part is at considerable distance from the rim of those remnants. Because the thin loamy coversands suitable for agriculture (code 23) occur exclusively on top of a till layer, these areas were possibly uninviting for settlement and the building of hunebeds. However there may, perhaps, have been another reason. Hunebeds were usually built less than 300-400m from the rim of the boulderclay,[12] which suggests that most boulders were found in a 400m wide zone along the rim of the boulder-clay remnants. Further away the erosion would have been stronger, or would at least have lasted longer, and the boulders might have sunken in potholes in the sandy subsoil, or have been removed by the streams.

Acknowledgements

The drawings were made by A. Visser and B. Donker. The director of the Soil Survey Institute, Wageningen, permitted us to use the concept soil maps 12E, 17W and 17E.

Notes

1. General 20th century spelling was 'Drente' until the idea that 'ancient is respectable' made the provincial government decide about 1965 that it should be 'Drenthe', which is now official. The h is not pronounced.

2. Cf. also Bakker 1979a, 1979b, 1980, 1982, in which some of the subjects dealt with here are discussed in more detail. The present publication continues the paths taken in some of the earlier studies.

3. NAP = Normaal Amsterdams Peil, which is identical to the German Normal Null, N.N. ('peil' = level).

4. There is no confirmation yet that TRB pottery distributions display the effects of 'territories'.

5. Omitted were the dubious or not precisely located tombs D6f, D5b, D9a, D13a, D13b.

6. These figures were derived from the map. Kaelas (1981) gives other figures: 290 tombs per 500 sq km (or 1 tomb per 1.7 sq km; the land surface was perhaps attained by substracting all marshland and other empty areas). Kaelas explains the extreme tomb density in this area where 'four-fifths of all the passage graves in Sweden are concentrated in an area of 500 sq km' not by a wealthy farming population on these fertile soils, but, because pollen diagrams hardly showed any cereal pollen, by a prosperous fisher-hunter population.

7. The foreign user of the Dutch soil map is referred to the colour-illustrated Major soils and soils regions in the Netherlands (De Bakker 1979). A multilingual glossary for the soil types given above was compiled from De Bakker's data by Bakker (1982, 103). The translations given above are the terms of Avery (1973) according to De Bakker (1979).

8. The number of tombs is slightly different from those mentioned in section 10.4 because some of the dubious ones mentioned in note 5 were also included.

9. Though there is a reasonably good agreement between the hunebed sites and the dry areas in fig. 10.1 (which is based on much less detailed maps) it is also conspicuous that these dry areas become smaller towards the north and the west, whereas the long northwestern tip of the plateau is presently without dry areas and known hunebeds. This last observation poses a problem: when there were indeed many more dry areas in this tip, why are there no hunebeds? Perhaps because the coversands are much thicker here and because the boulderclay was here less eroded than in the areas rich in hunebeds. The as yet unpublished relevant sheets of the geological map may clarify this point (and help to narrow down the possible 'rules for selecting good hunebed sites' for TRB man).

10. We transferred tombs D46 and D47 from (e) to (b) on the basis of the 1783 Hottinger map, the 1850-52 topographical map and the geological map. Other ridges may also have become unrecognisable by the heathland reclamations and further land use.

11. The exceptions known to us are hunebed D26 (sheet 12E), which has a type (a) position, and D16 (sheet 12W) in a type (e) position; in both cases the coversand layer (of the surrounding area!) is apparently just under 120cm. Hunebed D54a (sheet 17W) would lie, according to van Giffen's unpublished location map (c) on Hn23x-V, which seems highly improbable. We assume that its position should be shifted to where the concept soil map gives a small coversand ridge on top of code Hn21-VI, which means that the code for the ridge is Hd21-VII.

12. Cf. section 10.4. We did not actually substantiate this with statistics, but the impression is strong enough. The necessary procedure would be to enlarge the smaller scale appendix geological map sheet 17E used for fig. 10.4 to scale 1:50,000 and to compare an overlay with those of the relevant data in the 1:50,000 soil and geological maps of this sheet.

References

Aaby, B., 1986. Trees as anthropogenic indicators in regional pollen diagrams from eastern Denmark. In: K.-E. Behre (ed.), Anthropogenic Indicators in Pollen Diagrams. Rotterdam: Balthema, 73-93.

Andersen, S.T., 1970. The relative pollen productivity and pollen representation of North European trees, and correction factors for tree pollen spectra. Danmarks Geologiske Undersøgelse 2. Rk., Nr. 96.

Andersen, S.T., 1980. The relative pollen productivity of the common forest trees in the early Holocene in Denmark. Danmarks Geologiske Undersøgelse Årbog 1979, 5-19.

Andersen, S.T., 1985. Natur- og kulturlandskaber i Naesbyholm Storskov siden istiden. Fortidsinder 1985, 85-124.

Avery, W.B., 1973. Soil classification in the Soil Survey of England and Wales. Journal of Soil Science 24, 324-338.

De Bakker, H., 1979. Major soil and soil regions in the Netherlands. Wageningen: Centre for Agricultural Publication and Documentation; The Hague-Boston-London: W. Junk Publishers.

Bakker, J.A., 1976. On the possibility of reconstructing roads from the TRB period. Berichten van de Rijksdienst voor het Oudheidkundig Bodemonderzoek 26, 63-91.

Bakker, J.A., 1979a. The TRB West Group. Cingula 5. Amsterdam: A.E. van Giffen Instituut voor Prae- en Protohistorie.

Bakker, J.A., 1979b. Protection, acquisition, restoration, and the maintenance of the Dutch hunebeds since 1734: an active and often exemplary policy. Berichten van de Rijksdienst voor het Oudheidkundig Bodemonderzoek 29, 143-183.

Bakker, J.A., 1980. Einige Bemerkungen über die niederländischen Grossteingräber und deren Erbauer. Nachrichten aus Niedersachsens Urgeschichte 49, 31-59.

Bakker, J.A., 1982. TRB settlement patterns on the Dutch sandy soils. Analecta Praehistorica Leidensia 15, 87-124.

Behre, K.-E., K. Brandt, D. Kučan, P. Schmid and W.H. Zimmermann, 1982. Mit dem Spaten in der Vergangenheit, 5000 Jahre Siedlung und Wirtschaft im Elb-Weser-Dreieck. Wilhelmshaven: Landesinstitut für Marschen- und Wurtenforschung.

Booij, A.H. and M.F. van Oosten 1978. Bodemkaart van Nederland schaal 1:50,000. Toelichting bij de kaartbladen 17 West Emmen en 17 Oost Emmen. Wageningen: Stichting voor Bodemkartering.

Burrichter, E., 1976. Vegetationsräumliche und siedlungsgeschichtliche Beziehungen in der Westfälischen Bucht. Ein Beitrag zur Entwicklungsgeschichte der Kulturlandschaft. Abhandlungen aus dem Westfälischen Museum für Naturkunde 38 (1), 3-14.

Casparie, W.A. and W. Groenman-van Waateringe 1980. Palynological analyses of Dutch barrows. Palaeohistoria 22, 7-65.

Giffen, A.E. van, 1925, 1927. De Hunebedden in Nederland, vols. I (1925), II (1927) and Atlas (1925-1927). Utrecht: Oosthoek.

Groenman-van Waateringe, W., 1986. Grazing possibilities in the Neolithic of the Netherlands based on palynological data. In: K.-E. Behre (ed.), Anthropogenic Indicators in Pollen Diagrams. Rotterdam: Balkema, 187-202.

Harsema, O.H., 1982. Settlement site selection in Drenthe in later prehistoric times, criteria and considerations. Analecta Praehistorica Leidensia 15, 1982, 145-159.

Havinga, A.J., 1984. A 20-year experimental investigation into the differential corrosion susceptibility of pollen and spores in various soil types. Pollen et spores 26, 541-557.

Hegener, M., 1983. De hunebedden in oostelijk Drenthe en Renfrews territorial marker theorie. Unpublished thesis, Archaeological Institute University of Utrecht.

Huiskes, B., 1985. Van veldnaam tot vindplaats? Een onderzoek naar het verband tussen hunebedden an 'steennamen' in Drenthe. Driemaandelijkse Bladen, orgaan van het Nedersaksisch Instituut der Rijksuniversiteit te Groningen, 37, 81-94.

Iversen, J., 1964. Retrogressive development of a forest ecosystem demonstrated by pollen diagrams from fossil mor. Oikos supplement 12, 35-49.

Jager, S., 1984. Een reconstructie van een prehistorische weg in de gemeente Anloo (prov. Drenthe) alsmede eeuwenoude fossiele karresporen. Unpublished thesis Biologisch-Archaeologisch Instituut University of Groningen. English version to appear in Palaeohistoria.

Kaelas, L., 1981. Megaliths of the Funnel Beaker Culture in Germany and Scandinavia. In: Renfrew (ed.), 1981, 77-91.

Kahlke, H.D., 1981. Das Eiszeitalter. Leipzig-Jena-Berlin: Urania-Verlag.

Kinnes, I., 1975. Monumental function in British Neolithic burial practices. World Archeology 7, 16-29.

Klok, R.H.J., 1979. Hunebedden in Nederland, zorgen voor morgen. Haarlem: Fibula-Van Dishoeck.

Klok, R.H.J., 1982. Oude graven tussen IJssel en Vallei, prehistorische grafheuvels uit de steen- en bronstijd. Schaffelaarreeks 11. Barneveld: BDU.

Luijten, H., 1985. Onderzoek naar gelijkenisgroepen in het Nederlandse Trechterbekeraardewerk uitgaand van D26: een begin. Unpublished thesis IPP, Amsterdam University.

O'Kelly, M.J., 1982. Newgrange. Archaeology, Art and Legend. London: Thames and Hudson.

Picardt, J., 1660. Korte beschryvinge van eenige vergetene en verborgene antiquiteten ... tusschen de Noord-Zee, de Yssel, Emse en Lippe. Amsterdam: Gerrit van Goedesbergh.

Renfrew, C., 1973a. Before Civilization - The Radiocarbon Revolution and Prehistoric Europe. London: Jonathan Cape.

Renfrew, C., 1973b. Monuments, mobilization and social organization in neolithic Wessex. In: C. Renfrew (ed.), The Explanation of Culture Change, 539-558. London: Duckworth.

Renfrew, C., 1974. Beyond a subsistence economy: the evolution of social organisation in prehistoric Europe. In: C.B. Moore (ed.), Reconstructing Complex Societies, Supplement to the American School of Oriental Research 20 (Cambridge Mass.), 69-84.

Renfrew, C., 1976. Megaliths, territories and populations. Dissertationes Archaeologicae Gandenses 16, 198-220.

Renfrew, C., 1980. Toward a definition of context: the North German megaliths. Nachrichten aus Niedersachsens Urgeschichte 49, 3-20.

Renfrew, C., 1981. Introduction: the megalith builders of western Europe. In: C. Renfrew (ed.), 1981, 8-17.

Renfrew, C. (ed.), 1981. The Megalithic Monuments of Western Europe, the latest evidence presented by nine leading authorities. London: Thames and Hudson.

Schlicht, E., 1968. Die Funde aus dem Megalithgrab 2 von Emmeln, Kreis Meppen, Studien zur Keramik der Trichterbecherkultur im Gebiet zwischen Weser und Zuidersee. Göttinger Schriften zur Vor- und Frühgeschichte 9. Neumünster: Karl Wacholtz Verlag.

Schuldt, E., 1972. Die mecklenburgischen Megalithgräber. Berlin: VEB Deutscher Verlag der Wissenschaften.

Stolp, J., 1977. Bodemkaart van Nederland Schaal 1:50 000, Toelichting bij kaartblad 12 Oost Assen. Wageningen: Stichting voor Bodemkartering.

Taayke, E., 1985. Drie vernielde hunebedden in de gemeente Odoorn. Nieuwe Drentse Volksalmanak 1985, 125-144.

Tiesing, H., 1974. Harm Tiesing over landbouw en volksleven in Drenthe edited by C.H. Edelman. Vols. 1-2, 2nd ed. Assen: Van Gorcum and Comp. B.V. (1st ed. 1943).

Ter Wee, M.W., 1979. Toelichtingen bij de Geologische Kaart van Nederland 1:50 000, Blad Emmen West (17W) en Blad Emmen Oost (17O). Haarlem: Rijks Geologische Dienst.

Wieringa, J., 1968. Iets over de ligging van de hunebedden op het zuidelijk deel van de Hondsrug. Nieuwe Drentse Volksalmanak 1968, 149-157, Plates I-V.

11. SOILS OF PRE-HISPANIC FIELD SYSTEMS IN THE RIO SALADO BASIN, NORTHERN CHILE - A PRELIMINARY REPORT

Helen C.M. Keeley

Abstract

Agricultural terraces are one of the most widespread landscape features of the arid and semi-arid tropical and sub-tropical regions of the Andes. In northern Chile terrace cultivation is the main method of agricultural production in the Andean piedmont (2,500 - 4,000m above sea-level). Today most of these terraces lie abandoned and very little is known about past agricultural practices and economy in the area and their implications for archaeology and rural development. Recent research at Paniri (Late Intermediate) and other archaeological sites in the Rio Salado basin illustrates the diversity and complexity of agricultural systems in an arid environment.

11.1 Introduction

11.1.1 Location

The Rio Salado Basin is situated in the province of El Loa in the Second Region of Chile (Antofagasta), which is one of the driest places on earth. The Rio Salado is a tributary of the Rio Loa, which follows a major transverse valley reaching the sea from a source in the Andean Cordillera. The Loa is 400km long and follows a tortuous route to the coast. The Salado joins the Loa just south of Chiu-Chiu (fig. 11.1). The convergence of drainage in the big bend basin of the Loa river creates a series of oases, the largest of which is Calama. Sixteen kilometres north of Calama is the important open-cast copper mining centre of Chuquicamata.

The Loa has been defined as a heterogeneous channel resulting from the union of several independently originated sections (Mortimer 1980).

11.1.2 Geology and geomorphology

The principal geomorphic features of the area are: a) sedimentary sands in a sector of the Turi basin (the Turi plain), b) a rhyolitic plateau, descending from 3,400m in Caspana across 40km to 2,500m at Chiu-Chiu and c) the pre-Altiplanic ranges of the Andes. The most notable volcanic formations are Cerro Leon, Toconce and Paniri and the Cordillera del Chao - a lava flow between the mountains of Paniri and Leon which represents the largest single lava flow in the world.

Fig. 11.1 Geographical Location of the Rio Salado Basin in Northern Chile.

11.1.3 Climate

The climate is that of a marginal high desert; in the sub-Andean zone precipitation occurs sporadically in summer, while in the high Cordillera (Andean zone) precipitation is more regularly concentrated between January and March. Climatic stations are based at Ollague (3,700m): annual precipitation 70.7mm, mean annual temperature 6.8°C, and Calama (2,260m): annual rainfall 0.0mm, mean annual temperature 13.3°C. Records maintained at Toconce (3,300m) for 1976 and 1977 show that precipitation is concentrated in the summer months and monthly distribution of precipitation between years is variable (Villagran et al. 1981).

Extreme temperature gradients exist between night and day conditions. On the altiplano severe night frosts may occur at any time of the year down to -20°C but, in contrast, afternoon temperatures may rise to 28°C. The air is extremely dry and winds are desiccating (Wright 1963b).

11.1.4 Vegetation

The high rhyolitic lava platform, which forms part of the altiplano region of the Andes in northern Chile, supports open steppe vegetation, consisting of wiry tussock grasses and low xerophytic shrubs, which seldom achieves a closed plant cover (Wright 1963b). Thirty to 70% of the ground surface remains bare except for six to eight weeks in February and March when brief ephemeral plants emerge. The only close-growing permanent vegetation is that which develops wherever there is a supply of spring or river water, where cushion-bog communities ("bofedales") develop. The severe environmental regime has led to development of specialised vegetation: tussock grasses Stipa frigida and Stipa ichu are commonest, along with a xerophytic shrub of the genus Baccharis and a dwarf tree known as "Quenua" (Chuquiragua oppositifolia) (Wright 1963b).

Villagran et al. (1981) examined a vegetation transect from Turi (3,100m) to Cerro Leon (4,250m) across the Toconce area (22 degrees 06 minutes - 22 degrees 22 minutes South, 68 degrees 15 minutes - 68 degrees 07 minutes West) above the upper limit of the Atacama desert. Along the transect the total vegetation cover ranged from 2.5% at the lower end to 35.7% at 4,000m altitude. Four major physiognomic zones, including six associations, were recognised within the transect: 1) the 'Pre-Puna' belt (2,700-3,150m), characterised by a sparse cover of xerophytic subshrubs, 2) the 'Puna' belt (3,150-3,850m), physiognomically dominated by shrubs and presenting the highest vegetational cover, (this scrub vegetation is also known as 'Tolar'), 3) the 'High Andean' belt localised above the 'Puna', 4) the 'Subnival' belt, a high mountain semi-desert with a poorly developed vegetation; only three species occurred at the upper vegetational limit (4,380m).

An ethnobotanical study carried out in the Toconce area by Aldunate et al. (1983) examined the perception of the environment and use of flora by the inhabitants of this pre-Altiplanic community. Three units of landscape utilisation were distinguished, i.e. "Cerro" at higher elevations, which is of both ritual and economic significance (including gathering of Llareta - a cushion plant, used for fuel), "Campo", which is a pastoral area, mainly for llamas, and "Chacra", which consists of man-made terraces on the steep slopes of the Andean canyons, where crops are grown. The inhabitants of Toconce were found to recognise ecological zones which provide different resources, e.g. wood (Llareta) and medicinal plants are more abundant at higher elevations (Cerro), llamas graze freely in the Pajonal (upper Campo) but after the seasonal rains are herded down to lower zones in the Tolar, where a dense herbaceous cover develops. Agricultural labour is concentrated in the Chacra, which is contiguous to the village.

11.1.5 Soils

The soils of northern Chile have been described by Diaz Vial and Wright (1965). Arid zone soils occur from 18 to 30 degrees South, including the Andean altiplano with associated high mountain peaks, the Andean slopes and associated outlying foothills and the plains (pampas) from sea-level to about 2,300m. There are three readily distinguishable soil regions corresponding to three distinct environmental zones and nine separate soil assemblages have been differentiated.

Region 1 includes much of the northern part of the coastal range and adjacent plains to the east - saline soils predominate.

Region 2 occupies a low position on the broad Andean piedmont (500-2,300m). The soils - Typical Red Desert soils and related Lithosols and Regosols (FAO classification) - are alkaline rather than saline.

Region 3, including the upper piedmont, over the foothills and main Cordillera ranges to the Andean Altiplano, contains soils with more varied chemical characteristics and five distinct assemblages have been recognised (Diaz Vial and Wright 1965), ranging from true desert soils towards semi-desert, i.e. a) coastal desert soils, b) desert soils weakly impregnated with salts, c) Greyish and Brownish Sierozem-like soils of the cool desert of the Andean foothills and adjacent plains, d) Lithosols related to minimal Brown Forest soils and Sierozems; considerable development of recent colluvial and alluvial soils in part extensively terraced by former civilisations; some areas of Paramo-like soils, Non-Calcic Brown soils and Calcic Brown soils. This region extends from about 32 degrees S. north to the Peruvian border, occupying the higher slopes of the Andes between about 3,000 and 3,600m; e) steppe-like grassland soils of the Andean Altiplano.

The environmental characteristics of the soil assemblages most relevant to the Rio Salado basin are described in detail: (c) Assemblages dominated by Greyish and Brownish "Sierozem-like" soils of the Cool Upland Desert (map unit 7).

The landscape of this area comprises the piedmont plains, which are partly buried under old outwash fans and younger alluvial and colluvial materials, and ranges of hills along the foot of the Andean Cordillera, extending in places up to the very steep slopes of the Cordillera itself almost to the very limits of the Altiplano. It includes deeply incised cordilleran valleys, once famous for their terrace garden agriculture, and some very large salars, e.g. the Salar de Atacama.

Soils show some characteristics of desert soils but also some commonly associated with Sierozem and Calcic Brown soils. The oldest and most stable soils are those of the ancient piedmont remnants, which are most easily seen in the far north and are usually buried beneath younger alluvial and colluvial materials further to the south. Topsoil colours are pink to pale reddish brown becoming more reddish and yellowish with depth until (at about 30cm) a very reddish-yellow horizon with calcium carbonate ($CaCO3$) accumulation is reached. A moderate amount of $CaCO3$ occurs in the topsoil but strong accumulation is restricted to the lower subsoil. A small amount of clay is present throughout the profile. Topsoils show slightly to moderately developed vesicular structures but there are no salts in the profile. These stable soils occupy only a very small percentage of the area.

More common are related lithosols (developed on steeper cordillera slopes) and hill and steepland soils obviously related to the stable zonal soil. Hill and steepland soils are fairly shallow, calcareous and show characteristic yellowish to reddish-brown coloured topsoils. Many alluvial and colluvial fan soils occur, some similar to regosolic hot desert soils; others older and more stable, showing incipient zonal characteristics. Solonchak and "Takyr-like" soils occur in some of the larger basin depressions, often along the margins of the salars.

Weathering processes are operating weakly but about equally on topsoil and subsoil. Soil textures heavier than loams are rare but even the more sandy soils (apart from the youngest regosols) appear to contain some clay. There is no sign of extensive clay eluviation, although leaching is sufficient to remove some $CaCO3$ and concentrate it in the subsoil and, also, removes soluble salts from at least the upper 150cm but the salts reappear towards the toe of many alluvial fans, associated with zonal soils, and may form small local salars in the sloping plains. Processes associated with the organic regime are better developed than in most of the other true desert soils. Erosion is locally severe, mainly due to the great intensity of rainfall during summer showers. Wind erosion is very active where the natural plant cover is open.

(d) Assemblage dominated by Lithosols related to Minimal Brown Forest Soils (map unit 8):

The landscape is almost wholly mountainous with very steeply incised streams. In a few places at the heads of some of the larger valleys, terrace garden agriculture is practised at the foot of the mountain slopes.

Soils are Lithosols and steepland soils, the majority showing relationships with Minimal Brown Forest Soils. There is an increase in the intensity and depth of soil weathering through the range; many soils are developed from volcanic ash. Much of the terrace garden agriculture has now stopped but extensive grazing continues.

(e) Steppe grassland soils of the Andean Altiplano (map unit 9):

The landscape comprises the wide undulating altiplano, mountain ranges and associated colluvial and alluvial soils, local peat bogs and salars.

Soil variation exists between the drier south and the wetter north and is also related to the presence of volcanic ash. Soils are weakly weathered and slightly to moderately leached, developing under semi-arid, cold (extreme) climatic conditions. Agriculture consists of extensive grazing.

11.1.6 Archaeology

The regional chronology for the Upper-Middle Loa area has been summarised by Nunez (1978) and is shown in fig. 11.2. Cultures of the Formative Period - Vega Alta I and II and Loa I and II - all show intensive development of agriculture and animal breeding, along with proliferation of monochrome pottery. Lasana I and II is associated with extensive agriculture including keeping of camelids. Regional development in the upper Loa is characterised by the formation of defensive enclosures and settlements associated with pottery bearing a red wash, along with a strong monochrome tradition shown in Lasana II and San Pedro III. Dupont, near Calama, illustrates the last moments of the pre-Inca sequence. The local sequence was interrupted by settlements with chullpas of uncertain date but originating in the Altiplano.

Aldunate and Castro (1981) and Lynch (1983) have postulated integrated cultural development in the South-Central Andes centred on the drainage basin of Lake Titicaca, the Puna de Atacama being in the southern part of this area suggesting that the Chilean Norte Grande - from the Atacama desert northward - participated in this cultural area from before Tiwanaku times, as shown in fig. 11.3. The defining features of the South-Central area are pervasive aridity, vertical zonation of the environment, southern tricolour pottery and primacy or "nuclearity" of the circum-Titicaca altiplano. The Lake Titicaca basin has long been suspected as the centre of camelid domestication, representing the largest single high quality area for grazing in the Andes (Lynch 1983). It is generally considered that cultivated diploid potatoes arose from wild species native to the altiplano around Lake Titicaca; dung is required as a fertilizer for potatoes.

In the South-Central Andes from the dry puna and barren deserts of the altiplano down through deeply entrenched drainages to the irrigated coast, settlement has always been discontinuous and resource zones often sharply demarcated and shared by various peoples. Lynch (1983) considers that from an early pattern of transhumance by hunter-gatherers through pastoral nomadism and caravan networks to stylistically and economically integrated regional civilisation, cultural development in the South-Central Andes has an integrity of tradition that is dependant on camelids and their products.

Figure 11.2: Regional Chronology (Nunez 1978)

Peruvian	N. Chile	Upper-Middle Loa	AD/BC
Colonial	Colonial	Colonial	
			1500
Late Horizon	Inca	Inca	
			1400
Late Intermediate	Regional development	Dupont/Toconce Lasana II	
			1000
Middle Horizon	Tiwanaku		
		Lasana I	
			500
Early Intermediate	Regional Formative	Loa II	300
		Loa I	0
			200
Early Horizon		Vega Alta II Vega Alta I	800
Initial			
Preceramic periods	Hunter-gathering	Several complexes from Salado - Loa Talabre	2000
			5000
			8000
		Tuina	
			9000
			10,000

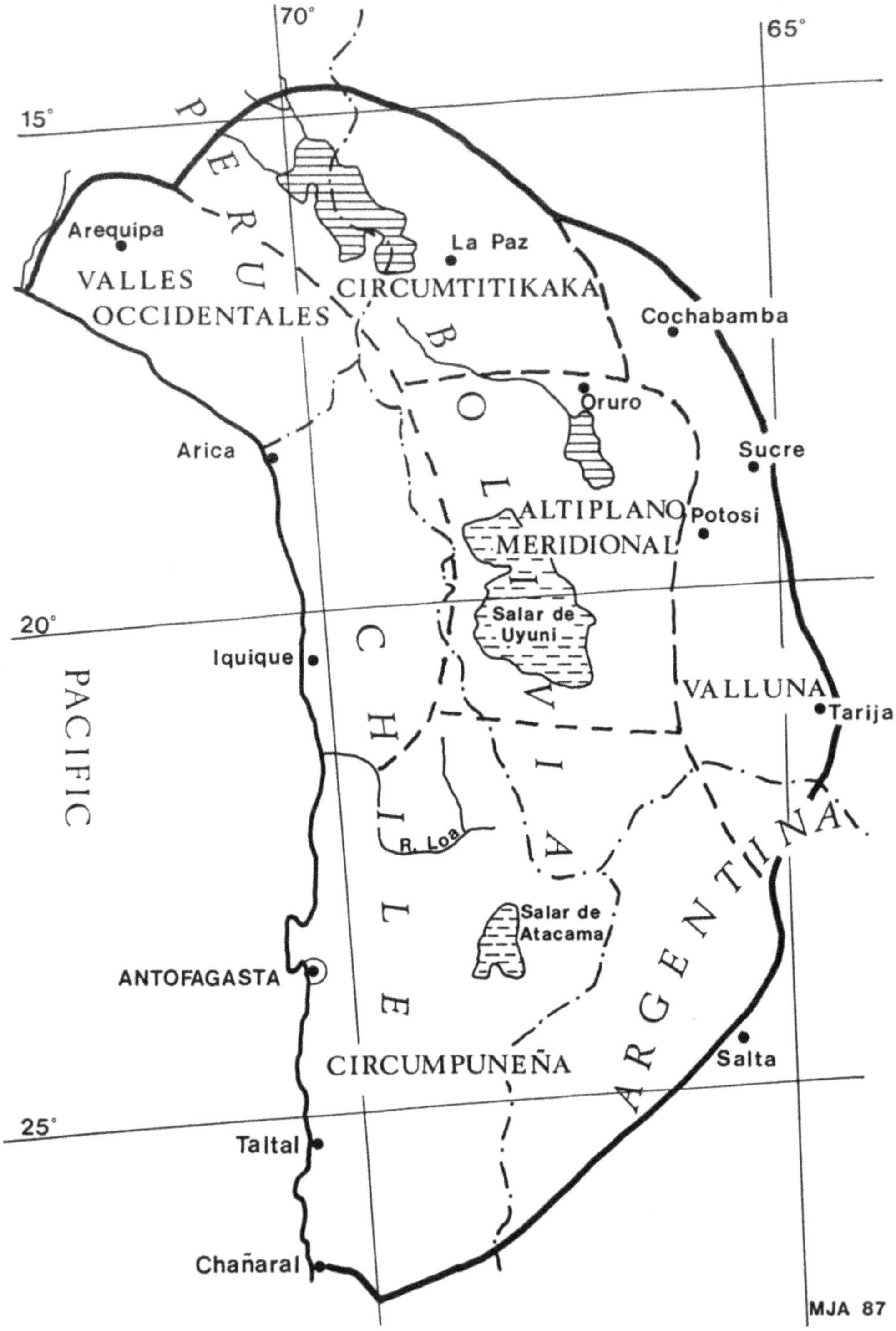

Fig. 11.3 The South-Central Andes Cultural Area.

In the headwaters of the Rio Loa there are suggestions of camelid domestication by about 200 bc and in Loa II (ad 100-400?) llama herding is confirmed by corrals and other archaeological evidence of sedentism, e.g. crop-raising, larger settlements and high frequency of locally made ceramics (Lynch 1983).

11.2 Agricultural field systems in the Rio Salado basin

A terrace may be defined as a constructed cropping surface with a slope less than normal to the site utilised (Field 1966). Terraces usually consist of stone retaining walls behind which soil is accumulated for crop growing. A terraced landscape represents a large labour investment and provides the basis for sustained agriculture over many generations. The agricultural terrace is an important kind of agricultural landform, one expression of human modification of the surface of the earth (Field 1966). Terracing is one of the important techniques man has developed and used in adapting the physical resources of the earth's sloping lands. Thus soils have been created, normal erosion has been delayed and diverted, hydrology has been altered and microclimates have been produced and changed.

Terraces allow the control of water. Even without supplementary irrigation, the increased soil depth normally affords more moisture to crops. Rainfall may accumulate and infiltrate to a greater extent on the flatter slope of a terrace and it is possible to create a soil profile that will either retain water or drain well, depending on the needs of the crop it would be designed for. With irrigation the terrace makes possible a nearby complete control of the vegetative-reproductive cycle. In areas subject to frost, the advantage of slope and exposure placement with respect to sunlight and air drainage is great. The technique fits well into an early point in the trend towards mass handling of crops. The size of most terraces falls into a range that is convenient to the work of small groups of people using hand tools. In modern agriculture the limited size of a terrace unit is normally a disadvantage, impeding the mechanised mass handling of standardised crops.

Many of the higher pre-Columbian cultures used terraces in the basin of Mexico, in Chiapas and in Guatemala. In South America the distribution extends, with gaps, from Santa Marta in Columbia to the Choapa River in Central Chile, on both faces of the Andes. The Incas arrived in northern Chile and Argentina late in their period of expansion. Most of the terraces were long established before the Inca conquest, according to Field (1966), although Donkin (1979) states that the remarkably uniform andenes at Toconce, Topain, Ayquina and Cupo suggest Inca influence. Toconce has the greatest display of intensive terracing with stylistic unity in Chile (Field 1966), thought to be the result of determined planning (Donkin 1979).

The soils of this region are generally thin and sites favourable for agriculture and terracing, e.g. outwash fans, are limited and widely scattered (Donkin 1979). The only deep soils are those formed from volcanic ash deposited directly on the high altiplano, certain colluvial soils formed partly from volcanic ash and rock waste accumulating in the bottom of quebradas and soils formed from volcanic material deposited as alluvium in the lake basin of the Salar de Atacama (Wright 1963a). At no time since man first appeared in the region has the climate been sufficiently humid to permit any considerable amount of soil weathering and, as a result, soil formation is minimal. The oldest layers of volcanic ash on the altiplano are loamy coarse sands. Very little loose soil is found in places on the rhyolite slope of the Cordillera and all is stony and sandy. There is abundant evidence throughout the region of a consistent trend in the direction of increasing aridity (Wright 1963a). According to Donkin (1979), more than 50% of all anciently terraced land in Chile and Argentina is now abandoned, for what are likely to be predominantly physical reasons - a very small reduction in average rainfall or greater fluctuations between years would be critical for agriculture in peripheral locations and there is also a strong possibility of gradual lowering of the ground water, leading to the disappearance of springs.

Water sources are generally inconveniently located in narrow gorges so that water must be taken off very high up. Water quality is often a problem due to high concentrations of carbonate and chloride salts.

The only ecological niches suitable for permanent dwellings are the quebradas with streams and springs. The development of terracing began with the deliberate construction of silt traps by cross walls across the quebradas. Construction of long lateral walls to exclude animals from the crops was carried out, thus creating lateral terraces (Wright 1963a). The simplest terrace system in valley bottoms comprises wide platforms of soil produced by alluvial siltation upstream of crude rubble barriers. Many are still in use and were later provided with irrigation canals (Wright 1963a). On these terraces the deeper layers of soil usually contain more humus than the upper layers; in some cases dark humus-rich subsoil extends to over 120cm. When still in use, the soils often have a rich worm population (<u>Allolobophora caliginosa</u>). In some cases the lowermost part of the subsoil contains finely divided charcoal - probably the result of burning off vegetation in the valley bottom and sides.

These terraces contrast strikingly with the tiers of well-constructed narrow bench terraces along the valley sides, fed by neat strong aqueducts, which represent a much more highly evolved system of soil management (Wright 1963a). The soils are lighter and more even in colour than most of the valley bottom soils and usually contain a higher proportion of stones and coarse sand, and much less silt and clay, in the upper part of the profile. Many show a distinct change at about 40 to 45cm, below which the soil is more dense and compact. The texture of the soil often changes from sandy loam or stony loam to stony clay loam at this point. Agricultural fields at a number of locations in the Rio Salado basin are described below.

11.2.1 Toconce

Toconce is located on a branch of the Rio Salado at an altitude of 3,350m and has 260 inhabitants. There is terracing up and down the stream, chiefly on the left bank and generally facing north. Only a small proportion of the fields are used now for cropping, being those easiest to irrigate and closest to the village. The main crops are broad beans, maize, alfalfa, vegetables and flowers. Herding is important and the animals graze in the Vegas of Turi and several other places (i.e. areas of pasture near sources of water). However, according to Aldunate et al. (1981), the importance of herding is decreasing and the terraces have become more important for production of crops for the consumer market of Calama.

In his reconnaisance of southern Andean agricultural terraces Field (1966) noted two types of terrace at Toconce:

(a) Small bench terraces comprising square-cornered flat field benches with downslope water drops at either end, approximately 1 to 1.5m wide, 4 to 6m long and about 70cm high at the front wall. Those terraces which were abandoned had lost at least 5cm of soil.

(b) Bench terrace series, which are similar to the first category but about three times the surface area and are the main ones still cultivated for maize and alfalfa.

The date of construction of these terraces is still uncertain but it is likely that the major phase coincided with agricultural expansion in the Late Intermediate period.

11.2.2 Ayquina

Ayquina is situated in the Rio Salado canyon at 2,980m and has a population of 176. Bench terraces, about 10 ha of which remain under cultivation, are irrigated by springs and water is stored in dammed basins called cochas. The main crops are alfalfa and maize, with some garlic, potatoes, vegetables and flowers, and llamas graze in the Vegas of Turi.

11.2.3 Caspana

At 3,260m in the valley of the Rio Caspana, the village has a population of 350 and approximately 35 ha of bench terraces under cultivation. Barley and wheat are important crops, in addition to maize and alfalfa, and fruit is grown, including apples, pears, apricots and plums.

11.2.4 Cupo

Cupo at an altitude of 3,650m has a population of about 80. Field (1966) recognised three types of terrace:

(a) Benches, which must have been built under unified direction.

(b) Irregular benches.

(c) Cross-slope levelings with bunds instead of stone walls.

Only about five ha are still cultivated for alfalfa, maize and some vegetables. Llamas and sheep are grazed in the Vegas of Paniri.

11.2.5 Turi and Lasana

These settlements do not have terraces but levelled fields for both pasture and food cropping, designed with the primary purpose of facilitating irrigation. Turi is on the sandy plain and supplied by a thermal spring. Most of the houses belong to people from Ayquina and Toconco who graze their animals here. Alfalfa and wheat are cultivated. Lasana, where there is a Pukara or fortified settlement of Late Intermediate date, is closer to Calama and is a flourishing market garden centre.

11.3 Archaeological investigations at Paniri - preliminary results

Paniri is at an altitude of 3,200m and was chosen for detailed study after a brief reconnaisance of the area in 1982. The work forms part of a much wider archaeological and ethnographical research project being carried out in the Upper Rio Loa region by Dra. Victoria Castro of the Department of Archaeology and Anthropology of the University of Chile and Dr. Carlos Aldunate, Director of the Museum of Precolombian Art, Santiago. The abandoned settlement is surrounded by a large area of agricultural fields (mainly terraces) and irrigation canals centred on the Quebrada Paniri. The site extends for about 1km along the quebrada and across about 3km at its widest point. The main aim of the project is to investigate how and over what period the agricultural system worked, the relationship of the settlement with others, on both a local and a regional basis, and the reasons for its abandonment. Little archaeological work has been carried out in the past at Paniri and there are no maps or aerial photographs available which are sufficiently detailed to provide a base-map for the site.

Relatively few fields, i.e. those in the immediate vicinity of the water supply, are still cultivated - some by the remaining occupant of the valley and the rest by people from other villages (Cupo, Turi, Ayquina and Lasana) who have rights to the land. Maize appears to be the predominant crop, followed by broad beans and alfalfa. Smaller areas are dedicated to sunflowers, household vegetables and potatoes and a fruit-bearing cactus known as 'tuna'. Fruit trees such as pears, peaches and plums are also grown. The water at Paniri, which comes from springs and the melting snows of Mount Paniri (5,960m), is known to be "sweet", whereas that at Cupo is associated with soil salinity problems, but it is not abundant and a rota system is observed for irrigation. At Paniri and Cupo llama seems to be the main source of meat, but only consumed on special occasions, and the wool is spun and knitted locally.

This report is based on the results of two short periods of field-work carried out in 1984 and 1985. Archaeological survey was begun in 1984 by Fernando Maldonado and Carlos Urrejola and continued in 1985 by Tim Malim and Debbie Simons, concentrating on the soil study areas. Also in 1985 Debbie Simons carried out an initial examination of the ancient settlement and Lucila Recart Haynes began a modern land use survey.

11.3.1 The irrigation system

The terraced fields received irrigation water from a sophisticated canal system supplied by water from the slopes of Mount Paniri and the upper Quebrada Paniri and investigation of this system was begun in 1985 by Tim Malim. A field walking survey of eight traverses crossing Quebrada Paniri at right angles was carried out and a total of 15 stone-sided canals was discovered, which are fed by springs regulated by at least eight reservoirs, which are all situated on the western, gentler slopes of the quebrada. Several stone aqueducts (four found so far) supplied water to the terraces on the eastern side and a major canal (Plate 11.I), hugging the cliff face and built up to some 4m above the valley bottom, swung round to enter the ancient settlement at Paniri. This canal may have flowed over the largest aqueduct (Plate 11.II) (2.6m high and 12.6m long) to supply a farther system of fields east of the settlement. However, it is still by no means clear from where water was taken to supply the highest terraced fields as few remains of canals are apparent.

One of the characteristics of Paniri, which has not been seen at any other sites in the area, is an abundance of stone mounds (Plate 11.III) which are so frequent and of such grand proportions and fine finish that it is tempting to suggest that they had additional functions to mere field clearance and were accessories to the irrigation system. Canals often run along the edges of the mounds and run from the mounds into the fields. It may be that the presence of the mounds reduced evaporation from the water channels but further work is required to understand their role.

11.3.2 The fields and associated soils

Three main types of fields have been noted:

(a) Levelled bench terraces on the valley sides (Plate 11.IV), (b) shallow terraces on gentler slopes (Plate 11.V) and (c) cross channel terraces in the valley bottom (Plate 11.VI). As is it generally considered that the development of terracing begins with the deliberate construction of silt traps by cross walls across the quebradas, it is most likely that the cross channel terraces (c) are the oldest. The

Plate 11.I The Main Canal near the Ancient Settlement at Paniri.

Plate 11.II Aqueduct adjacent to the Ancient Settlement at Paniri.

Plate 11.III Stone Mounds at Paniri.

Plate 11.IV Bench Terraces at Paniri.

Plate 11.V Shallow Terraces on a Gentle Slope at Paniri.

Plate 11.VI Cross Channel Terraces in the Quebrada Paniri.

bench terraces along the valley sides and the small fields on the interfluves, fed by canals and aqueducts, represent a much more highly evolved system of soil and water management and thus (a) and (b) would have developed later than (c). Pottery collected from the terrace surfaces and from excavated profiles indicated a Late Intermediate date, corresponding to the Toconce period of 850 to 1100 AD, for the main occupation at Paniri but some earlier elements were present and the site was probably occupied from at least the Middle Horizon period because of the good quality of the water supply.

The soils of five areas of abandoned terraces have been examined and sampled to date and analyses, including particle size distribution, organic matter content, pH and levels of nutrient elements, are in progress. It is also intended to carry out phytolith analysis on selected profiles. Four representative soil profiles are shown in fig. 11.4 and described below:

S II T2A

The profile represents one of a series of bench terraces situated on the eastern side of the Quebrada Paniri, northeast of the ancient settlement. The slope was approximately 25 degrees and the terrace was on the south side of a small irrigation canal. Vegetation consisted of small herb and shrubs, including Malva lechera and "Tolar", and cacti. The terrace was 3m long, 2m wide and the front and back walls were 50cm high. The surface of the terrace was flat and on the surface were a few sherds of pot and many small stones. The north-facing section, from the back to the front of the terrace, is described:

0 to 8cm	Dark brown (10YR3/3) coarse sand washed down from upslope.
8 to 46cm	Sloping boundary; dark yellowish brown (10YR4/6) friable coarse sandy loam with weak medium sub-angular blocky structure. Stones 20% gravel to large; roots few coarse to fine fibrous and woody.
Below 46cm	Dark yellowish brown (10YR3/4) structureless, friable coarse loamy sand. Stones 50% increasing with depth to 80%, gravel to large (including large boulders); roots few fine fibrous. Bedrock was reached at 80cm.

S III T1B

This section was dug through one of a small group of cross-channel terraces on the other side of the quebrada, to the west of the previous profile. The terrace was south-facing, on gently sloping (5 degrees) land and appeared to have been surrounded by canals on three sides. It had a flat surface and was 10 x 2m. The fields had not been recently cultivated and supported various herbs and grasses. Little surface pot was found. The soil was developed on gravels of the river bed and the surface was very gravelly. There was a 40cm drop from the top of the front wall to the terrace below. The east-facing section from the back to the front of the terrace is described:

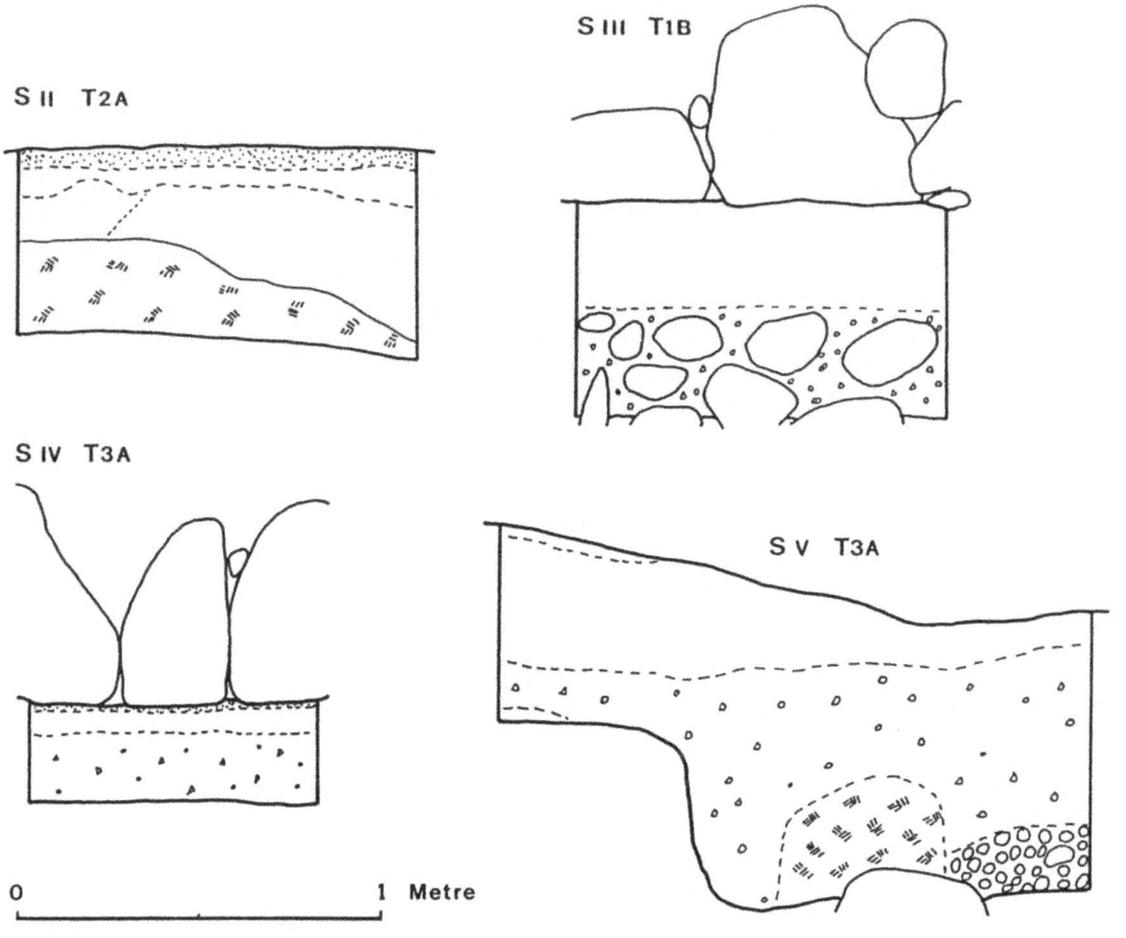

Fig. 11.4 Terrace Soil Profiles at Paniri.

0 to 30cm	Very dark greyish brown (10YR3/2) friable coarse gravelly humic loamy sand with weak subangular blocky structure. Stones 20% gravel to medium (mainly gravel); roots abundant coarse to fine fibrous and woody.
Below 30cm	Gravels and boulders of the river bed: very dark grey (10YR3/1) friable structureless gravelly coarse sand matrix containing rounded and angular pebbles and some very large boulders. Roots common coarse to fine fibrous and woody. The base of the pit was at 1m.

All the sides of the pit were similar and the profile was little modified by terracing.

S IV T3A

A series of terraces to the south of the settlement was examined, west of the Quebrada Paniri, and these seemed to be very eroded - with perhaps 30 to 60cm of soil lost. The area is bounded by a wall to the west (no doubt built to keep animals out of the crops), below which terraces cover the steeper slopes and flatter fields are associated with many canals. Many stone mounds, which also occur elsewhere on the site, occupy this area. The slope of the land was about three degrees, although the field surfaces were generally flat, and various herbs grew. The terraces were about 4 x 2m and that described below was bounded by a canal to the east.

0 to 7cm	Dark brown (10YR3/3) friable coarse loamy sand with weak medium subangular blocky structure. Roots common coarse to fine fibrous; stones 10% gravel to medium.
7 to 25cm	Dark yellowish brown (10YR3/4) friable coarse loamy sand with weak medium subangular blocky structure. Stones 20% gravel to medium; roots common coarse to fine fibrous and woody.
Below 25cm	Large boulders with matrix similar to above.

S V T3A

An area of rather broken-down terraces was examined, to the south of a large stone mound. The slope of the land was approximately 21 degrees although the terrace surfaces were fairly flat. A representative field was 4 x 2m, with walls 50cm at back and front and a canal to the south. The north-facing section from the back to the front of the terrace is described at the mid-point:

0 to 15cm	Dark greyish brown (10YR4/2) coarse loamy sand. Roots common coarse to fine fibrous and woody; stones 10% gravel to medium (mainly gravel).

15 to 80cm	Yellowish brown (10YR5/6) coarse loamy sand. Roots common coarse to fine fibrous and woody; stones increased from 10% to 50% with depth (gravel to medium).
Below 80cm	Indurated material (calcrete), which had been deliberately dug out at the front of the terrace, presumably to allow the depth of soil to be increased and to facilitate movement of water downslope from terrace to terrace.

In general soil pH was slightly alkaline, ranging from 6.9 to 9.2 with a mean of 8.3, and soil conductivity was low, ranging from 0.5 to 6.5 with a mean of 1.7, giving no indication of salinity at Paniri.

The fields on gently sloping land were very small (usually 4 x 2m across), with shallow drops to the fields below and bounded by canals on one or both sides. The soils were very shallow and contained very little fine material, being generally coarse loamy sand and rarely coarse sandy loam in texture. They appeared to be extremely eroded by both wind and water action, no doubt due to lack of a vegetation cover, with perhaps 30 to 60cm of soil lost. This was deduced by looking at the present soil surface in relation to the tops of the stone walls and canals to which, in the past, the soil would have reached.

The soils of valley side bench terraces had clearly been eroded but were still much deeper than on the flatter site, reaching depths of up to 1m. In some cases in the lower part of the profile darker material, containing more clay than the overlying material, occurred - similar to that noted by Wright (1963a) in the Salar de Atacama region. The soil is deeper at the front of the terrace than at the back in order to provide a level surface for cultivation and, in some cases, the terrace builders had to cut through thick layers of calcrete to provide the required soil depth and facilitate drainage downslope.

The cross-channel terraces in the valley bottom were generally much larger than the other terraces but this may not have been the case in the past. It is these terraces which are predominantly cultivated now and what were originally small fields may have been amalgamated to facilitate modern cultivation (wheat has been an important crop in recent years). Those which had fallen into disuse appeared to have been considerably eroded. Adjacent to the Vega, on the other side of the quebrada from the ancient settlement, some of the valley soils were very calcareous and had a very high water table due to water seeping from springs at the base of the slope. These soils would probably be included in Melendez and Wright's (1962) Paniri Association of gleyed and half-bog soils derived from recent alluvium but field conductivity readings indicated that they are not saline. Although containing more fine material than any of the other soils examined, they are still dominated by sand. The soils which are still being cultivated are almost completely remade periodically and soil is brought from nearby uncultivated areas, if needed, to maintain the level in relation to the field walls.

11.4 Comments

The Incas appear to have had very little impact on this area - apart from a large Inca building within the Late Intermediate settlement at Turi, there is very little evidence of their presence. Various reasons have been put forward for terrace abandonment and population decline since the time of first contact with the Spaniards in the 16th century. These include decimation and relocation of the population under the influence of European diseases and colonial mining, re-orientation of the agricultural economy to pastoralism, introduction of European domestic animals - leading to over-grazing - and increasing aridity of the climate. The argument for climatic change, however, requires verification.

In the upper Loa foothills industrial development and accelerated growth of the neighbouring urban centres of Calama and Chuquicamata has had a strong impact on the cultural system. Early in this century the Llareta was massively exploited as fuel for the copper mining industry at Chuquicamata. The towns are attractive, especially to the young, resulting in a breakdown of cultural patterns. Restrictions on border crossings have interrupted the social contact and commercial interchange with the neighbouring Andean villages in Bolivia and Argentina. The enormous need for water in Calama, Chuquicamata and Antofagasta has resulted in a shortage for the Andean villages and consequent reduction in carrying capacity for livestock and the area of cultivated land.

Clearly prior to the Spanish conquest Paniri was a flourishing agricultural settlement, as were other centres such as Toconce, Cupo, Turi and Ayquina. The sophistication of the field and irrigation systems implies unified direction in their construction and thus a relatively high degree of social organisation. Priorities for future work include production of a basic map of Paniri, continuation of the irrigation survey, investigation of the stone mounds, further studies of the development of the agricultural fields (in particular to obtain dating evidence for different phases of field construction) and archaeobotanical studies (particularly phytolith analysis) to try to find out what crops were grown in the past, combined with the modern land use survey. Such information should provide a basis for comparison with other sites in the region.

11.5 Conclusions

1. The Rio Salado basin is a marginal high desert area, with characteristically limited vegetation, dominated by three main soil assemblages, i.e. Greyish and Brownish Sierozem-like soils of the Cool Upland Desert, Lithosols related to Minimal Brown Forest soils and Steppe grassland soils of the Andean Altiplano.

2. Remains of pre-Inca settlements with associated fields and irrigation canals are quite numerous and some, such as at Toconce and Paniri, have highly sophisticated terrace systems which would have required unified direction in their construction and thus a relatively high degree of social organisation.

3. The large area of agricultural fields (mainly terraces) at Paniri received water from a sophisticated canal system supplied by water from the slopes of Mount Paniri and the upper Quebrada Paniri. Spring water was regulated by reservoirs and transported by stone-sided canals and aqueducts. Numerous stone mounds found on the site may have been associated with the irrigation system.

4. There are three types of field at Paniri, i.e. levelled bench terraces on the valley sides, shallow terraces on gentler slopes and cross-channel terraces in the valley bottom. It is likely that the latter were the earliest while the others represent a more highly evolved system of soil and water management and correspond to the main period of occupation, dated from pottery to the Late Intermediate (Toconce) period of 850 to 1100 AD.

5. Soils were generally of slightly alkaline pH (mean 8.3) but showed no signs of salinity (mean conductivity 1.7). All soils were sandy to varying degrees and also, in many cases, extremely stony. All appeared to have been eroded by both wind and water action.

6. The soils of the small fields on gentle slopes and interfluves were very shallow and the most extremely eroded, appearing to have lost 30 to 60cm of soil.

7. Valley side soils also appeared to have been eroded but were generally quite deep (up to 1m), particularly at the front of the terraces where soil had been built up behind the wall in order to provide a flat surface for cultivation.

8. The cross-channel terraces were generally the largest and soils had developed on river alluvium, much of which was very stony except in the waterlogged areas adjacent to the Vega.

9. The major phase of terrace construction in the Rio Salado basin appears to have coincided with agricultural expansion in the Late Intermediate period, during which time it is thought that the Chilean Norte Grande participated in a South-Central Andean cultural area centred on the drainage basin of Lake Titicaca.

10. The reasons for abandonment of the terraces are still uncertain but it now seems likely that depopulation resulting from the Spanish conquest was a major factor. This decline is continuing due to modern pressures on the rural cultural system in the upper Rio Loa region.

Acknowledgements

I would like to thank Carlos Aldunate, Victoria Castro and Carlos Urrejola for providing the opportunity for me to work with them in the north of Chile (Project: "Sistemas de Asentamiento en la Region del Loa Superior: Patrones Arqueologicos y Etnograficos", grants DIB S1435-8533 and S1435-8544). I am most grateful to the Royal Society, the British Academy, the Wenner-Gren Foundation, the University of London Central Research Fund and the Royal Geographical Society for providing grants for this research.

I would like to thank all those who have participated in the work, particularly Fernando Maldonado, Pilar Alliende, Tim Malim, Debbie Simons, Lucila Recart Haynes, Lindsey O'Callaghan, Ron and Mary Jarvis. I am very grateful to those organisations and people who have provided information and assistance - there are too many to name all but I would like to mention the Department of Earth Sciences of the Open University (who allowed me to hire their Landrover in Chile), Wye College of Agriculture (laboratory analyses), the Museo Chileno de Arte Precolombino in Santiago, the British Geological Survey, Warwick Bray, Bente Bittmann, and our neighbour at Paniri (Sra. Julia Cruz) who made bread for us.

References

Aldunate del Solar, C. and Castro Rojas, V., 1981. Las Chullpa de Toconce y su relacion con el poblamiento altiplanico en el Loa superior periodo tardio. Santiago: Kultrun.

Aldunate, C., Armesto, J.J., Castro, V. and Villagran, C., 1983. Ethnobotany of a pre-Altiplanic community in the Andes of northern Chile. Economic Botany 37(1), 120-135.

Diaz Vial, C. and Wright, C., 1965. Soils of the arid zone of Chile. Soils Bulletin 1, Rome: FAO.

Donkin, R.A., 1979. Agricultural terracing in the aboriginal New World. Viking Fund Publications in Anthropology 56, University of Arizona Press.

Field, C., 1966. A reconnaissance of southern Andean agricultural terracing. Unpublished Ph.D. thesis, University of California.

Lynch, T.F., 1983. Camelid pastoralism and the emergence of Tiwanaku civilisation in the South-Central Andes. World Archaeology 15 (1), 1-14.

Melendez, E. and Wright, C., 1962. Los suelos de la region central del valle del Loa y su desarollo agricola. Unpublished report for IREN: TAM/RE/194.

Mortimer, C., 1980. Drainage evolution in the Atacama Desert of northern Chile. Revista Geologica de Chile 11, 3-28.

Nunez, A.L., 1978. Northern Chile. In: Taylor, R.E. and Meighan, C.W. (eds.), Chronologies in New World archaeology, 483-511. London: Academic Press.

Villagran, C., Armesto, J.J. and Arroyo, M.T.K., 1981. Vegetation in a high Andean transect between Turi and Cerro Leon in northern Chile. Vegetatio 48, 3-16.

Wright, A.C.S., 1963a. The soil process and the evolution of agriculture in northern Chile. Pacific Viewpoint 4, 65-74.

Wright, A.C.S., 1963b. "Los Bofedales" - alkaline cushion-bog peats of the semi-arid Chilean Altiplano. Pacific Viewpoint 4, 189-193.

12. THE ROLE OF GEOLOGY IN THE ARCHAEOLOGICAL FIELD SURVEY IN THE OER-LJ ESTUARY (NETHERLANDS)

P.C. Vos and R.W. Brandt

Abstract

In the coastal areas of the Netherlands, a close relationship exists between the occupation history and the geological development. Therefore, geology is of great importance to field surveys in these areas. Geology can supply important information on locational choices by (pre)historical inhabitants of certain regions. Also, geology is of importance to the survey method to be employed. In this paper, the role of geology with respect to the survey strategy of three areas in the former Oer-IJ estuary will be discussed.

12.1 Introduction

The Oer-IJ estuary ("Oer" in Dutch means "old, former") lay north-west of Amsterdam (fig. 12.1). During the Bronze Age and the Iron Age it was connected with the North Sea and during the prehistory the estuary was an isolated region because it was cut off from the rest of the coastal area by large peat areas; it was accessible only via the coastal barriers and dunes area, and via the small rivers draining the peat areas (fig. 12.2).

The occupation history of the Oer-IJ region was closely related to the geological developments in the estuary. The coastal barriers and dunes area, and the peat area lay so high that they were not flooded, not even during extremely high-water levels (fig. 12.3). Therefore, the sandy soils of the coastal barriers and dunes area have been occupied continuously since their development (in the area of the Oer-IJ estuary from ca. 4000 BP). In contrast, in certain parts of the peat area, where peat formation continued, drainage was so poor that occupation was impossible.

Within the estuary, during periods of strong marine influences, the intertidal and supratidal areas were not suited to occupation. Only during low tides was the intertidal area suited for short visits by man. The supratidal area could be visited safely for longer periods because it was flooded only during spring-tides and storm surges. During periods in which the estuary was completely or partly isolated from the sea, the intertidal and supratidal areas were not flooded anymore. Then these parts of the estuary were suited to prolonged occupation, provided that the drainage capacity of the area was sufficient. From the Early Middle Ages onwards the Oer-IJ estuary was conclusively isolated from the sea by new coastal barriers (Jelgersma

Fig. 12.1 Location of the study area.

Fig. 12.2 Palaeogeographic reconstruction of the "Oer IJ" estuary, ca. 400 BC.

1. coastal barriers covered by dunes
2. idem., presumed
3. beaches and other sandy deposits between the coastal barriers, where there is a possibility of peat formation in the depressions.
4. estuarine deposits: tidal flats and salt marshes
5. reedy, fen peat
6. oligotrophic raised bog
7. tidal inlet
8. large tidal channels
9. border of the Velserbroek Polder, Uitgeesterbroek Polder and Assendelver Polders
   - I Velserbroek Polder
   - II Assendelver Polders
   - III Uitgeesterbroek Polder
10. schematic cross-section with paleo-tidal ranges

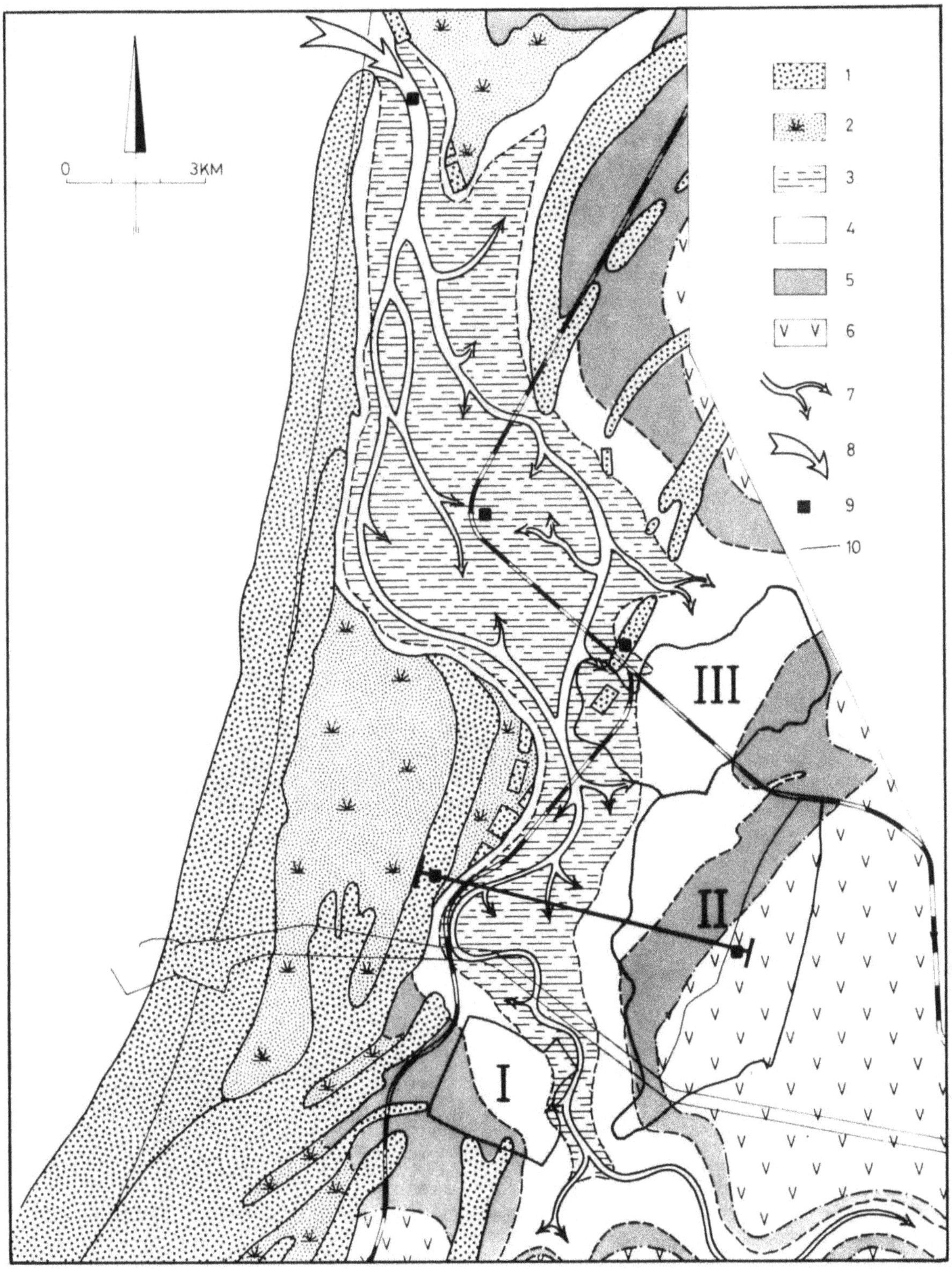

Fig. 12.2 Palaeogeographic reconstruction of the "Oer IJ" estuary, ca. 400 BC.

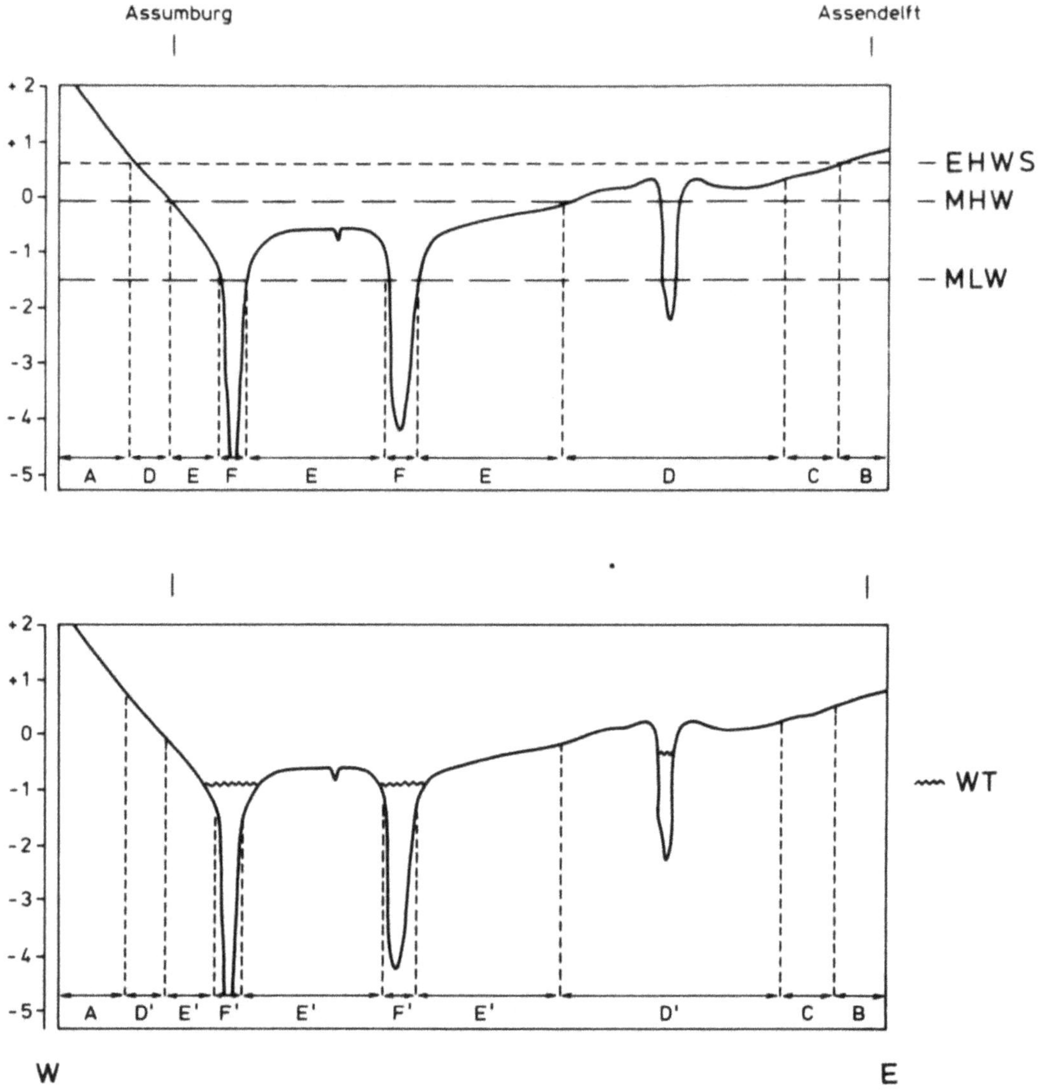

Zone above EHWS : A coastal barriers and dunes
B raised bog

Supertidal zone : C reedy fen peat
D tidal-levees and basins

Intertidal zone : E clay and sandy tidal flats

Subtidal zone : F main tidal channels

EHWS : extreme high-water-spring-tide

MHW : mean high water

MLW : mean low water

WT : water table

E' : former intertidal zone

Fig. 12.3 Schematic cross-section with paleo-tidal ranges (location, see fig. 12.2).

et al. 1970). Therefore the drainage capacity in the area decreased markedly, so that large scale peat formation occurred and shallow lakes were created in the lower parts. Until the large scale peat reclamations (Besteman and Guiran 1983) from AD 1000 onwards, the area of the Oer-IJ estuary (with the exception of the coastal barriers and dunes area) was not suited to occupation.

The destruction of prehistoric sites as a consequence of the lowering of the water tables in large scale reallotments (Assendelver Polders and Uitgeesterbroek Polder) and town development (Velserbroek Polder) was the direct cause of the regional archaeological investigations of settlements (Brandt 1983) which have been carried out since 1979 by the Albert Egges van Giffen Institute voor Prae en Protohistorie (IPP). The investigation started in the Assendelver Polders (1979-1981). Because in the late seventies the reallotment process in this area was far advanced, it was necessary to set up a large-scale investigative campaign in which the field survey had to coincide with the excavations. During this campaign much attention was paid to the geological development of this area. The knowledge of the geological development and the occupation history, obtained during this campaign of investigation, has strongly influenced the set-up of the field survey in the other parts of the Oer-IJ region.

For two reasons geology is of importance to field surveys:

1. Since there exists a relationship between the geological development in an area and the occupation history, geology can supply important information concerning locations where one has a good chance of finding traces of habitation, or - in the case of erosion - little chance of finding habitation layers;

2. Occupation layers may be covered with younger deposits; the thickness of these deposits is of great importance with respect to the survey method to be employed.

Another factor that is important to the strategy of the field survey is the land use (grass land or arable land) (Bult 1983). The area of the former Oer-IJ estuary mainly consists of grass land. Therefore we will leave out of consideration the differences between field survey in grass lands and in arable lands.

In this paper we will go into the role that geology has played in the determination of the investigation strategy of the field surveys which were carried out in the Uitgeesterbroek Polder in 1983 and 1984, and in the Velserbroek Polder in 1984 and 1985.

In Section 2, we will discuss the relation between the geological development and the occupation history in the Assendelver Polders (because this relation was of great importance to the strategy of the field survey in the other areas of the estuary) and review the methods which were employed during the field survey in these Polders.

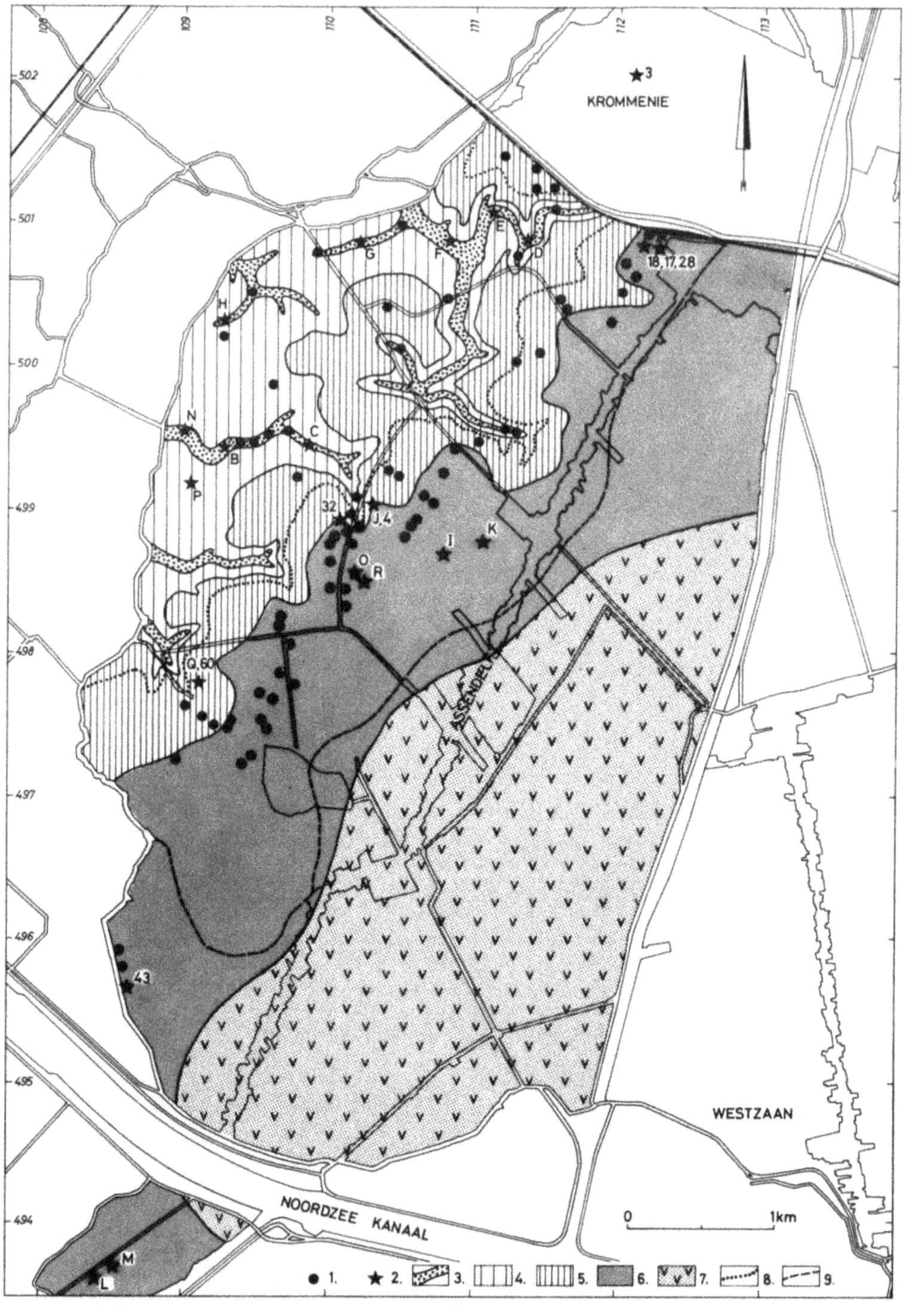

Fig. 12.4 Geological map of the Assendelver Polders. The map represents the clastic deposits and peat layers from the period of the Dunkirk 0 and Dunkirk I transgression phases (1500 - 0 BC). The Dunkirk 0 and Dunkirk I deposits are covered by a medieval clay layer of 30 - 40cm. In the geological map this layer is excluded.
1. sites surveyed, 2. sites excavated, 3. tidal creek and tidal levee deposits, 4. lagoonal and backswamp deposits (more than 100cm thick), 5. back-levee marsh deposits (less than 100cm thick), 6. reedy fen peat, 7. oligotrophic peat, 8. maximal extension of reedy fen peat on backswamp deposits, 9. maximal extension of oligotrophic peat (older than the Dunkirk 0 transgression).

In Section 3, we will deal with the field surveys of the Uitgeesterbroek Polder and the Velserbroek Polder separately: because these areas have experienced a considerably different geological development, the strategy of the field survey had to be different as well.

12.1 The Area of the Assendelver Polders

12.2.1 Geological development and distribution pattern of settlements

Fig. 12.4 represents the geological situation of the Assendelver Polders during the late Iron Age. This geological map shows that the polders can be subdivided into two areas:

1. the area with clastic deposits (mainly clay),

2. the peat area.

Within the area with clastic deposits ("clay area"), the following sedimentary environments can be distinguished: creeks, tidal levees, backswamps (supratidal), and tidal flats (intertidal). The peat area can be subdivided into *Phragmites* peat and oligotrophic peat.

In the geological map, the sites[1] that were found during the field survey by amateur archaeologists (Helderman 1965; 1971) and by the IPP are indicated by dots. The excavated sites are indicated by stars. The distribution pattern of the sites shows that the settlements from the late Iron Age are mainly located in the tidal levees and the border zone of the peat area, but also in the backswamps.

The geological development and its relation with the occupation history of the Assendelver Polders during the late Bronze Age and the Iron Age (Vos 1983; 1985) are summarized in Table 12.1.

Table 12.1: Landscape reconstruction and its relation with the occupation history of the Assendelver Polders

Landscape reconstruction	Occupation history
1000 - 1500 AD	
Around 1000 AD the Assendelver Polders were colonized again, when large scale drainage systems were constructed. Via the canals the sea could exert indirect influence on the area; as a result a clay layer of 25 to 50cm was deposited.	Permanent settlements to date
200 - 1000 AD	
The drainage of the area had become so bad that large-scale peat growth occurred in the former salt-marshes.	No permanent settlements

0 - 200 AD	
The old tidal creek system silted up. Therefore the drainage capacity in the area decreased. The inhabitants tried to keep up the drainage capacity by digging ditches. Probably the terps, about 20cm high, were thrown up in order to keep the houses dry during extreme rainfall in winter, when the fields in the former salt marsh were flooded.	Permanent settlements in former salt-marshes and border zone of the peat area
150 BC - 0	
The process of reduction of the tidal amplitude and extreme high water level had gone so far that the salt marshes in the estuary were not flooded anymore.	Permanent settlements in the former salt-marshes and border zone of the peat area
300 - 150 BC	
Because of the sedimentation the salt marshes silted up, and therefore these areas were flooded less frequently during spring-tides. The tidal amplitude and the extreme high water level gradually decreased because the tidal prism of the estuary (and consequently the tidal inlet) became smaller. The reduction of the tidal prism occurred because sedimentation was greater than erosion in the estuary.	No permanent settlement
550 - 300 BC	
Because of the further enlargement of the tidal inlet, the tidal amplitude increased and the extreme high water level rose. Because of the lowering of the peat surface by compaction and oxidation (caused by drainage) and the raised extreme high water level, the border zone of the peat was flooded.	No permanent settlements
650 - 550 BC	
The tidal influence increased further in the estuary. On the old tidal lagoon clay layers, tidal levees and backswamps developed. The tidal creeks drained the border zone of the peat area; consequently the peat growth stopped in this area.	Permanent settlements in the border zone of the peat area

700 - 650 BC

Because of an enlargement of the tidal inlet, the tidal amplitude, the extreme high water level, and the tidal currents increased in the estuary. As a result, a channel and creek system developed in the lagoonal clay area. This new system improved the drainage capacity of the border zone of the peat area. During this transition oligotrophic peat developed locally in this zone.	No permanent settlements

1000 - 700 BC

In the clay area a tidal lagoon developed, which was linked with the sea by a small tidal inlet. The water in the lagoon was brackish to fresh. In the whole peat area the peat growth continued.	No permanent settlements

12.2.2 Field survey

In the Assendelver Polders the following field survey methods were employed:

- Ditch-side survey. In autumn farmers are obliged (by the authorities) to clear their ditches. If a ditch cuts a site, archeological material may be dug up and deposited in the ditch sides during the clearing process.

- Mole-hill survey. Moles may push up archaeological material and therefore they are important to grassland field surveys.

- Relief mapping in the field. Prehistoric terps are observable in the field as small round elevations if they are ca. 20cm high. They can best be observed during autumn and winter, because in these seasons the grass is short.

- Aerial photograph research. Prehistoric traces (ditches and terps) in the shallow underground are observable in aerial photographs on the basis of crop marks.

- Core survey. The core hole survey is carried out by means of a "tree hole" drill and an auger. A "tree hole" drill is a tractor-driven drill which is used to dig holes (90cm deep and with 70cm in diameter) in which trees are planted. The auger (diameter 6 or 2cm) is pushed into the ground manually to a depth of max. 6-8m.

The core survey by means of the tree hole drill was applied very frequently in the clay area of the Assendelver Polders. Drawbacks of this method are: (a) it is labour-intensive compared to the other methods, and (b) not all the farmers will allow an intensive drilling program on their land. Other methods which were applied and which led to good results were relief mapping in the field and the use of aerial photography. The mediaeval clay layer (25-50cm) has not levelled the prehistoric relief because this layer covers the late Iron Age occupation level like a blanket. In the aerial photographs the terps can easily be recognized by the circular ditches around the terps, which have darker shades of grey than the other parts of the grassland. Besides the relief mapping and the aerial photograph research methods, ditch-side survey is an effective method to track down sites if they are cut by ditches.

In the peat area ditch-side survey is the most successful method. There are two explanations for this fact: firstly, because the other methods fail (see below), and secondly because in the peat area there are more ditches per hectare then in the clay area so that there is a better chance of cutting a site. Relief mapping in the field does not yield any results because no terps have been thrown up in the peat area. Also aerial photograph research does not produce satisfactory results because the ditches are filled up with peat so that in the aerial photographs no contrasts are observable. Core survey with the tree-hole drill is not possible as the farmers will not allow this, since depressions would develop after the filling up of the holes as a consequence of peat oxidation. Core survey with an auger (diameter 6cm) would be allowed, but because little material is brought up by this drill many borings would have to be made and even so, the results would be rather poor. This is because sites in the peat can only be recognized by sherds and wooden structures. Occupation layers which consist of organic material are hard to recognize in the auger, and bones have mostly been dissolved by the acid peat.

In the peat area a considerable number of mole-hills occur. During the ditch-side survey these mole-hills were also investigated. These investigations frequently led to the discovery of sherds. The number of sites which were discovered with the help of this method is small relative to the total number.

12.3 Field survey in the Uitgeesterbroek Polder and the Velserbroek Polder

12.3.1 Strategy of the field survey

The field survey, as carried out in the Uitgeesterbroek Polder and in the Velserbroek Polder, can be subdivided into four phases (fig. 12.5):

I compiling existing data,

II collecting new data,

III checking archaeological data per site,

IV advising on cultural resource management policy.

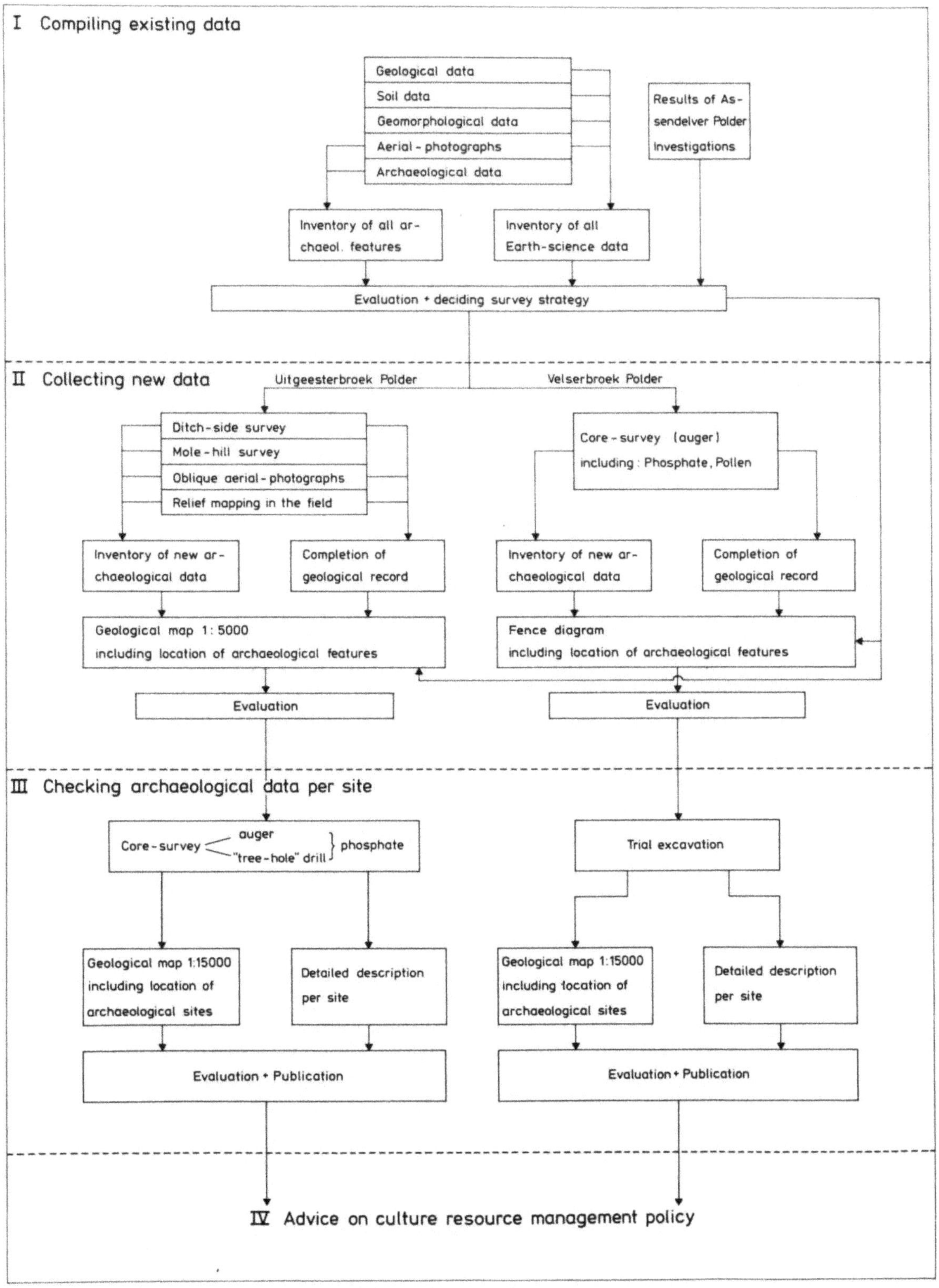

Fig. 12.5 Schematic representation of the investigation strategy of the field survey of the Uitgeesterbroek Polder and Velserbroek Polder.

During phase I, geological and archaeological data of each area were collected, and inventories of these data drawn up. These inventories, together with the results of the investigations in the Assendelver Polders, were the base on which the survey strategy of the next phase was determined.

12.3.2 The field survey in the Uitgeesterbroek Polder

Phase 1: compiling existing data. The geological data which were collected in the first phase consisted of geological, morphological and soil maps,[2] and a number of publications about the area (de Roo 1953; Pons and Kloosterhuis 1955; Zagwijn 1971). The archaeological data included excavation reports (Groenman-van Waateringe et al. 1961; Halbertsma 1964; Stuurman 1978) and the results of the field survey by amateur archaeologists (Helderman 1965; 1971). In addition, aerial photographs supplied valuable information concerning creek patterns and old ditch patterns.

After an inventory of the geological data was drawn, it appeared that the geological structure of the Uitgeesterbroek Polder was very similar to the structure of the Assendelver Polders. Just like the Assendelver Polders, the Uitgeesterbroek Polder consisted of a peat area and a clay area during the Bronze Age and Iron Age.

On the basis of the geological structure the following distribution pattern of prehistoric settlements could be expected in the Uitgeesterbroek Polders:

- early Iron Age, in the border zone of the peat area;

- late Iron Age, the greatest density of settlements in the tidal levees and in the border zone of the peat area, and - less densely - in the backswamps.

The drawing up of an inventory of the archaeological data revealed that most sites (20) appeared to lie in the border zone of the peat area. Only a few finds were made in the clay area (4). All the finds dated from the late Iron Age. After drawing up the inventory of the geological and archaeological data, one proceeded with determining the strategy of the next phase of the field survey: collecting new data.

On the basis of the experiences in the Assendelver Polders, it was decided to employ the following field survey methods in the clay area: ditch-side survey, mole-hill survey, aerial photograph research, and relief mapping in the field. Ditch-side survey, mole-hill survey and relief mapping can be performed simultaneously by an experienced archaeologist. During bad weather he can work out his data and carry out aerial photograph research. For phase II of the field survey, in particular W. de Vries-Metz took detailed oblique aerial photographs of the clay area.

Core survey with an auger or "tree hole" drill was not carried out in this phase of the investigation, because this method is very labour-intensive, and would (on the basis of the experience of the field survey of the Assendelver Polders) not add many sites to those that had already been found with the help of other methods.

The field survey methods that were applied in the peat area are analogous to those used in the Assendelver Polders: ditch-side survey and mole-hill survey.

Phase II: collecting new data. The ditch-side survey, mole-hill survey, oblique aerial photograph research, and relief mapping in the field were carried out in 1983 and 1984 by C. Ravesloot of IPP.

Besides archaeological data, the field survey also yielded new geological data. On the basis of the finds from the material thrown up along the ditches, the boundary between the clay area and the peat area could be determined accurately, and with the help of aerial photographs and relief mapping in the field the creek pattern and the tidal levees pattern could be determined in fair detail. Using these data and the geological data which were collected during phase I, a provisional geological map (scale 1:5,000) of the area was made, on which the (assumed) archaeological sites were also indicated.

During the evaluation, the strategy for the next phase of the field survey, checking archaeological data of sites, was determined.

Phase III: checking archaeological data per site. It was decided to investigate all the (assumed) sites with an auger or the "tree hole" drill. During this core survey the presence of an occupation layer, a terp layer, and archaeological material (quantitative and qualitative) were especially looked for.

In the clay area half of the sites were investigated with the "tree hole" drill, the remaining sites with an auger. In the peat area an auger was used on all the sites. During the core survey phosphate analyses were also made on the occupation layers.

A detailed description was made of each site in which the geological structure, the relief and shape of the settlement, the height of the terp layer and/or occupation layer, the archaeological material (quantitative and qualitative) and the concentration of phosphate (Eidt 1973) were mentioned.

A geological map (scale 1:15,000) of the area was also drawn, in which the locations of all the archaeological sites were indicated.

In total, 121 new sites were discovered during the field survey; 47% of them were located in the former tidal levees, 26% in the border zone, 20% in the former intertidal area, and 7% in the former backswamp (Brandt and Ravesloot 1985). All sites date from the late Iron Age. A single find from the early Iron Age was made in a tidal levee, indicating occasional visits to the area.

During the evaluation of phase III, the way one could best prevent the loss of the most valuable sites because of reallotments was considered. In general, loss of valuable sites can be prevented either by appealing to the Monuments Law or by excavating the sites. The results of the survey, the geological map, and the advice to the authorities to protect the most valuable sites, were included in the final report.

Phase IV: advice on cultural resource management policy. It was proposed to the authorities that 12 areas should be protected against the consequences of the reallotment by assigning to them the status of "archaeological monument". For four valuable sites, which could no longer be saved because of the construction of new ditches and roads, financial support from the provincial government was applied for. Subsidies were allocated to three sites, and these sites were excavated in the summers of 1984 and 1985.

12.3.3 Field survey in the Velserbroek Polder

Phase I: compiling existing data. The geological data of the Velserbroek Polder were rather limited; for instance, no detailed geological or soil maps exist of this area. However, a number of publications were available (Güray 1952; van Straaten 1957; Bennema and Pons 1957; Jelgersma 1970; Zagwijn 1971). It appeared from them that the transition zone between the coastal barriers/dunes area and the salt-marshes of the Oer-IJ estuary was located in the Velserbroek Polder. Archaeological sites in the area around the Velserbroek Polder were discovered by amateur archaeologists (Woltering 1979) during excavations that took place during several pipe-line laying projects in the area. Field survey techniques such as ditch-side survey yielded hardly any results.

In 1978, during the laying of a pipe-line, a number of barrows dating from the Bronze Age were discovered. Two barrows were excavated by the Rijksdienst voor Oudheidkundig Bodemonderzoek (Woltering 1979).

The finds proved that the Velserbroek Polder and its direct surroundings have been continuously occupied since the early Bronze Age. The largest number of finds was made in the transition area between the dunes and the salt-marshes. For prehistoric man, the transition area was the most favourable location to settle; both the estuary and the dunes area could be used for subsistence. The salt-marshes could serve as pastures for livestock. During extremely high water levels (e.g. storm surges), the dunes served as a shelter for the livestock (this in contrast to the situation in the Assendelver Polders). The transition zone from the salt-marshes to the dunes, which was flooded only incidentally, was favourable for agriculture because of the natural silt layers, which were deposited periodically, and the good soil composition. The more elevated sandy soils were much less fertile.

In the Velserbroek Polder the prehistoric finds were almost exclusively made during excavations; this can be explained by the fact that the prehistoric cultural layers lay deeper than 80cm under the surface. They are covered with dune sand and/or clay layers, consequently prehistoric groundtracks could not be recognized in the aerial photographs nor in the field. Also the ditch-side survey yielded hardly any results because the prehistoric layers lay too deep.

Since these methods were not very successful, it was decided to carry out a core survey with an auger in phase II of the field survey. The diameter of the auger had to be 2cm because it was almost impossible to push an auger with a diameter of 6cm into the sandy deposits. Moreover, because of the lack of cohesion of the sandy deposits, it would have been very difficult to bring up the material with an auger of 6cm diameter.

Phase II: collecting new data. The core survey in the Velserbroek Polder was carried out by W. Bosman in 1984 and 1985. In total, about 250 borings were made in the transition zone between the dunes area and the estuary. A detailed description of the sediment composition and the archaeological material was made from each boring. Also, in a large number of borings phosphate analyses and palynological investigations were made. The palynological investigation was directed only at pollen indicating agriculture in the direct surroundings, e.g. Cerealia and *Plantago lanceolata*.

The borings were plotted in a fence diagram, in which the locations of the archaeological finds were indicated as well. The fence diagram provides a fine view of the spatial distribution pattern of the dunes and the estuarine deposits, and it clearly delineates the relation between the geological structure and the distribution of the various culture layers.

As a check of the core survey, a control excavation was made in an important site in 1985.

Phase III: checking archaeological data per site. From the control excavation, it appeared that the geological record, which was based on the borings, was accurate (Bosman 1985). Besides, the control excavation proved that cultural layers were also present. In total, eight agricultural layers were excavated at depths varying between 1 and 2.50m below the surface (Beemster and Brandt 1985). The agricultural layers were separated by small dunes and estuarine humic clay layers. The oldest level dated from the middle Bronze Age. At present, the excavation results are being analysed further and a geological map (scale 1:15,000) of the Velserbroek Polder is being drawn.

Phase IV: advice on culture resource management policy. Since the future of the Velserbroek Polder had already been determined (a new quarter of the municipality of Velsen) it was too late to assign the status of "archaeological monuments" to parts of the polder. At present the IPP is working out an excavation scheme in order to prevent the irreversible loss of the most valuable sites.

12.4 Conclusions

- Making use of geological information in the field survey of coastal areas yields more archaeological sites than not doing so.

- Since the geological structure differs per area, it is impossible to develop a generally applicable strategy for a field survey in the coastal area. Each plot requires a separate approach.

- The field survey should be carried out before the future (e.g. reallotment) of the area is decided on. Then it is possible to protect or to excavate a number of important sites. Unfortunately, in practice (e.g. the Velserbroek Polder) the field survey is carried out after the determination of the future of the area, so that a lot of archaeological material is wasted.

Notes

1. A site is any location in which traces of (pre)historic habitation, artefacts, ecofacts, etc., or combinations of them have been found during the field survey.

2. Geological map, 1:50,000, blad 19 west. RGD, in press.
   Soil map of De Roo (1953) and Pons and Kloosterhuis (1955).
   Morphological map, 1:50,000, blad 19 west, STIBOCA.

Acknowledgements

We would like to thank W. Bosman and C. Ravesloot for their valuable advice on aspects of the surveys in the Velserbroek Polder and the Uitgeesterbroek Polder. We are grateful to J.T.E.M. Smit for translating the manuscript.

References

Beemster, D.C.M. and Brandt, R.W., 1985. Velserbroek Polder-Rugbyveld Hofgeesterweg HOG I. Intern rapport IPP.

Bennema, J. and Pons, L.J., 1957. The Holocene deposits in the surroundings of Velsen and their relations to those in the excavation. In: L.M.J.U. van Straaten and J.D. de Jong (eds.), The excavation at Velsen. A detailed study of Upper-Pleistocene and Holocene stratigraphy. Verh. Kon. Ned. Geol. Mijnbouwk. Gen., Geol. Ser. 17.

Besteman, J. and Guiran, A.J., 1983. Een middeleeuws-archeologisch onderzoek in Assendelft; een vroege veenontginning in middeleeuws Kennemerland. In: R.W. Brandt, G.J. van der Horst, and J.J. Stolp (eds.), de Zaanstreek archeologisch bekeken, 96-128. Zaanstad.

Bosman, W., 1985. Voorlopig verslag van de boorcampagnes 1983-1985 in de Velserbroek Polder. Intern rapport IPP.

Brandt, R.W., 1983. De Archeologie van de Zaanstreek. In: R.W. Brandt, G.J. van der Horst, and J.J. Stolp (eds.), de Zaanstreek archeologisch bekeken, 96-128. Zaanstad.

Brandt, R.W. and Ravesloot, C., 1985. De veldverkenning in de Uitgeesterbroek en Krommeniër Woudpolder. In: P.J. Woltering (ed.), Holland 17, 335.

Bult, E.J., 1983. Midden-Delfland, een archeologische kartering. Inventarisatie, waardering en bewoningsgeschiedenis. Nederlandse Archeologische Rapporten II.

Eidt, R.C., 1973. A rapid chemical test for archeological site surveying. American Antiquity 38, 206-210.

Groenman-van Waateringe, W., Glasbergen, W. and Hamburger, M.F., 1961. Een boerderij uit de eerste eeuw na Chr. te Krommmenie (N.H.). In: In het voetspoor van A.E. van Giffen, 110-128. Groningen: J.B. Wolters.

Güray, A.R., 1952. De bodemgesteldheid van de IJpolder. Boor en Spade 5, 1-28.

Halbertsma, H., 1964. Assendelft. NKNOB, 100-101.

Helderman, E.J., 1965. Tien jaar Zaanse archeologie. Noord-Holland, 124-129.

Helderman, E.J., 1971. Enige resultaten van vijftien jaar archeologisch onderzoek in de Zaanstreek. Westerheem 20, 52-83.

Jelgersma, S., Jong, J. de, Zagwijn, W.H. and Regteren Altena, J.F., 1970. Depositional history and coastal development in the western Netherlands; geology, vegetational history and archaeology. Mededel. Rijks Geologische Dienst, N.S. 21, 93-167.

Pons, L.J. and Kloosterhuis, J.L., 1955. Rapport van de verkenning van de bodemgesteldheid van het zuidelijk gedeelte van de provincie Noord-Holland. STIBOKA, rapport 409.

Roo, H.C. de, 1953. De bodemgesteldheid van Kennermerland-Noord. De bodemkartering van Nederland, deel XIV. Versl. Landbouwk. Onderz. 59-3. Wageningen.

Straaten, L.M.J.U. van, 1957. The Holocene deposits. In: L.M.J.U. van Straaten and J.D. de Jong (eds.), The excavation at Velsen. A detailed study of Upper-Pleistocene and Holocene stratigraphy. Verh. Kon. Ned. Geol. Mijnbouwk. Gen. Geol. Ser. 17, 93-218.

Stuurman, H.W., 1978. Een chaotische inheemse vindplaats uit de Romeinse Keizertijd in Assendelft. Westerheem 27, 81-105.

Vos, P.C., 1983. De relatie tussen de geologische ontwikkeling en de bewoningsgeschiedenis in de Assendelver Polders vanaf 1000 v. Chr. In: R.W. Brandt, G.J. van der Horst, and J.J. Stolp (eds.), de Zaanstreek archeologisch bekeken, 96-128. Zaanstad.

Vos, P.C., and Brandt, R.W., 1985. Publicatie Assendelver Polder Papers II. Rapport IPP, nr 85-376, Amsterdam.

Woltering, , 1979. Velsen Hofgeesterweg. Holland 11, 252-256.

Zagwijn, W.H., 1971. De ontwikkeling van het "Oer-IJ" estuarium en zijn omgeving. Westerheem 20, 11-18.

www.ingramcontent.com/pod-product-compliance
Lightning Source LLC
Chambersburg PA
CBHW040138240926
52019CB00019B/932
9780860545293